REPUBLIC OF SOUTH AFRICA
REPUBLIEK VAN SUID-AFRIKA

DEPARTMENT OF AGRICULTURE AND WATER SUPPLY
DEPARTEMENT VAN LANDBOU EN WATERVOORSIENING

FLORA OF SOUTHERN AFRICA

VOLUME 18

PART 3

ISBN 0 621 09698 9

FLORA OF SOUTHERN AFRICA

which deals with the territories of

SOUTH AFRICA, CISKEI, TRANSKEI, LESOTHO, SWAZILAND, BOPHUTHA-TSWANA, SOUTH WEST AFRICA/NAMIBIA, BOTSWANA AND VENDA

VOLUME 18

PART 3

SIMAROUBACEAE–MALPIGHIACEAE

by

K. L. Immelman, J. J. A. van der Walt, F. White &
B. T. Styles and P. D. de Villiers & D. J. Botha

Edited by

O. A. Leistner

Editorial Committee: B. de Winter, D. J. B. Killick, G. E. Gibbs Russell and O. A. Leistner

Botanical Research Institute,
Department of Agriculture and Water Supply

1986

CONTENTS

NEW COMBINATION PUBLISHED IN PART 3*

Sphedamnocarpus pruriens *(Juss.) Szyszyl.* subsp. **galphimiifolius** *(Juss.) P. D. de Villiers & D. J. Botha,* comb. nov., p. 3: 67

---

* Date of publication: November, 1986

INTRODUCTION

The Flora of southern Africa is arranged on the lines of the Engler system. Sequence and numbering of genera are as far as possible in agreement with De Dalla Torre & Harms (Genera Siphonogamarum, 1900 – 1907). Keys to families are provided in R. A. Dyer's Genera of Southern African Plants.

This part was compiled in accordance with a Guide to Contributors to the Flora of southern Africa (Ross, Leistner & De Winter, 1977). The latest edition of the Guide is available from the Librarian, Botanical Research Institute, Private Bag X101, Pretoria, 0001.

The following condensed abbreviations for literature references are used:

Burtt Davy, Fl. Trans.............	Manual of the Flowering Plants and Ferns of the Transvaal and Swaziland, Vol. 1 (1926) and Vol. 2 (1932)
C.F.A..............................	Conspectus Florae Angolensis
F.C...............................	Flora Capensis
F.C.B.............................	Flore du Congo et du Rwanda-Burundi
F.M.	Flora de Moçambique
F.S.W.A.	Prodromus einer Flora von Südwestafrika
F.T.A.............................	Flora of Tropical Africa
F.T.E.A.	Flora of Tropical East Africa
F.W.T.A.	Flora of West Tropical Africa
F.Z.	Flora Zambesiaca
R.A. Dyer, Gen.	The Genera of Southern African Flowering Plants by R. A. Dyer, Vol. 1 (1975) and Vol. 2 (1976).

Cited voucher specimens are all housed in PRE (National Herbarium, Pretoria).

Vol. 18 of the Flora, of which the present publication is a component, will appear in parts (see p. ix). The number of the part, namely 3, precedes the page number on all pages marked with Arabic numerals. This was done with a view to binding the entire volume, once completed, and to compiling a combined index to all its component parts. When binding the entire volume the pages marked with Roman numerals may be omitted.

The '3' preceding the page number of pages marked with Arabic numerals was inadvertently omitted from all text pages. Allowance will be made for this omission when a combined index to Vol. 18 is compiled.

PLAN OF FLORA OF SOUTHERN AFRICA

Crytogam volumes will in future not be numbered but will be known by the name of the group they cover. The number assigned to the volume on Charophyta therefore becomes redundant.

Alien families are marked with an asterisk.

Published volumes and parts are shown in italics.

Plase note that local prices as given below do not include GST. Prices given for other countries include postage.

INTRODUCTORY VOLUMES

The genera of Southern African flowering plants
    Vol. 1: *Dicotyledons* (Published 1975). Price: R11,23. Other countries: R14,00
    Vol. 2: *Monocotyledons* (Published 1976). Price: R8,21. Other countries: R10,00
Botanical exploration of Southern Africa (Published 1981). Price R40,00 (Obtainable from booksellers)

CRYPTOGAM VOLUMES

Charophyta (Published as Vol. 9 in 1978). Price: R4,25. Other countries: R5,30

Bryophyta:

    Part 1: Mosses: Fascicle 1: *Sphagnaceae— Grimmiaceae* (Published 1981). Price: R24,34. Other countries R30,40
                    Fascicle 2: Gigaspermaceae—Bartramiaceae (in press)
                    Fascicle 3: Erpodiaceae—Hookeriaceae
                    Fascicle 4: Fabroniaceae—Polytrichaceae

Pteridophyta (in press)

FLOWERING PLANTS VOLUMES

Vol. 1: *Stangeriaceae, Zamiaceae, Podocarpaceae, Pinaceae*, Cupressaceae, Welwitschiaceae, Typhaceae, Zosteraceae, Potamogetonaceae, Ruppiaceae, Zannichelliaceae, Najadaceae, Aponogetonaceae, Juncaginaceae, Alismataceae, Hydrocharitaceae* (Published 1966). Price: R1,98. Other countries R2,60

Vol. 2: Poaceae

Vol. 3: Cyperaceae, Arecaceae, Araceae, Lemnaceae, Flagellariaceae

Vol. 4: Part 1: Restionaceae
       Part 2: *Xyridaceae, Eriocaulaceae, Commelinaceae, Pontederiaceae, Juncaceae* (Published 1985). Price: R7,50. Other countries: R9,40

Vol. 5: Liliaceae, Agavaceae

Vol. 6: Haemodoraceae, Amaryllidaceae, Hypoxidaceae, Tecophilaeaceae, Velloziaceae, Dioscoreaceae

Vol. 7: Iridaceae: Part 1: Nivenioideae, Iridoideae
             Part 2: Ixioideae: Fascicle 1
                          Fascicle 2: *Syringodea, Romulea* (Published 1983). Price: R3,90. Other countries: R5,00

Vol. 8: Musaceae, Strelitziaceae, Zingiberaceae, Cannaceae*, Burmanniaceae, Orchidaceae

Vol. 9: Casuarinaceae*, Piperaceae, Salicaceae, Myricaceae, Fagaceae*, Ulmaceae, Moraceae, Cannabaceae*, Urticaceae, Proteaceae

Vol. 10: Part 1: *Loranthaceae, Viscaceae* (Published 1979). Price: R4,34. Other countries: R5,40
Santalaceae, Grubbiaceae, Opiliaceae, Olacaceae, Balanophoraceae, Aristolochiaceae, Rafflesiaceae, Hydnoraceae, Polygonaceae, Chenopodiaceae, Amaranthaceae, Nyctaginaceae

Vol. 11: Phytolaccaceae, Aizoaceae, Mesembryanthemaceae

Vol. 12: Portulacaceae, Basellaceae, Caryophyllaceae, Illecebraceae, Cabombaceae, Nymphaeaceae, Ceratophyllaceae, Ranunculaceae, Menispermaceae, Annonaceae, Trimeniaceae, Lauraceae, Hernandiaceae, Papaveraceae, Fumariaceae

Vol. 13: *Brassicaceae, Capparaceae, Resedaceae, Moringaceae, Droseraceae, Roridulaceae, Podostemaceae, Hydrostachyaceae* (Published 1970). Price: R10,00. Other countries: R12,00

Vol. 14: *Crassulaceae* (Published 1985). Price: R16,40. Other countries: R20,50

Vol. 15: Vahliaceae, Montiniaceae, Escalloniaceae, Pittosporaceae, Cunoniaceae, Myrothamnaceae, Bruniaceae, Hamamelidadeae, Rosaceae, Connaraceae

Vol. 16: Fabaceae: Part 1: *Mimosoideae* (Published 1975). Price: R13,59. Other countries: R16,75
                 Part 2: *Caesalpinioideae* (Published 1977). Price: R16,04. Other countries: R20,00
       Papilionoideae

Vol. 17: Geraniaceae, Oxalidaceae

Vol. 18: Part 1: Linaceae, Erythroxylaceae, Zygophyllaceae, Balanitaceae
         Part 2: Rutaceae
         Part 3: *Simaroubaceae, Burseraceae, Ptaeroxylaceae, Meliaceae* (Aitoniaceae), *Malpighiaceae* (Published 1986). Price: R9,65. Other countries: R12,00

Vol. 19: Polygalaceae, Dichapetalaceae, Euphorbiaceae, Callitrichaceae, Buxaceae, Anacardiaceae, Aquifoliaceae

Vol. 20: Celastraceae, Icacinaceae, Sapindaceae, Melianthaceae, Greyiaceae, Balsaminaceae, Rhamnaceae, Vitaceae

Vol. 21: Part 1: *Tiliaceae,* (Published 1984). Price R4,30. Other countries: R5,00
         Malvaceae, Bombacaceae, Sterculiaceae

Vol. 22: *Ochnaceae, Clusiaceae, Elatinaceae, Frankeniaceae, Tamaricaceae, Canellaceae, Violaceae, Flacourtiaceae, Turneraceae, Passifloraceae, Achariaceae, Loasaceae, Begoniaceae, Cactaceae* (Published 1976). Price: R8,68. Other countries: R10,75

Vol. 23: Geissolomaceae, Penaeaceae, Oliniaceae, Thymelaeaceae, Lythraceae, Lecythidaceae

Vol. 24: Rhizophoraceae, Combretaceae, Myrtaceae, Melastomataceae, Onagraceae, Trapaceae, Haloragaceae, Gunneraceae, Araliaceae, Apiaceae, Cornaceae

Vol. 25: Ericaceae

Vol. 26: *Myrsinaceae, Primulaceae, Plumbaginaceae, Sapotaceae, Ebenaceae, Oleaceae, Salvadoraceae, Loganiaceae, Gentianaceae, Apocynaceae* (Published 1963). Price: R4,53. Other countries: R5,75

Vol. 27: Part 1: Periplocaceae, Asclepiadaceae (Microloma–Xysmalobium)
         Part 2: Asclepiadaceae (Schizoglossum–Woodia)
         Part 3: Asclepiadaceae (Asclepias–Anisotoma)
         Part 4: *Asclepiadaceae (Brachystelma–Riocreuxia)* (Published 1980). Price: R4,43. Other countries: R6,00

         Asclepiadaceae (remaining genera)

Vol. 28: Part 1: Cuscutaceae, Convolvulaceae
         Part 2: Hydrophyllaceae, Boraginaceae
         Part 3: Stilbaceae, Verbenaceae
         Part 4: *Lamiaceae,* (Published 1985). Price: R22,00, plus G.S.T. Other countries: R28,00
         Part 5: Solanaceae, Retziaceae

Vol. 29: Scrophulariaceae

Vol. 30: Bignoniaceae, Pedaliaceae, Martyniaceae, Orobanchaceae, Gesneriaceae, Lentibulariaceae, Acanthaceae, Myoporaceae

Vol. 31: Part 1: Fascicle 1: Plantaginaceae, Rubiaceae, (Rubioideae—first part)
                 Fascicle 2: *Rubiaceae (Rubioideae*—second part): *Paederieae, Anthospermeae, Rubieae* (Published 1986). Price: R14,85. Other countries: R17,80

                 Fascicle 3: Ixoroideae, Chinchonoideae

         Part 2: Valerianaceae, Dipsacaceae, Cucurbitaceae

Vol. 32: Campanulaceae, Sphenocleaceae, Lobeliaceae, Goodeniaceae

Vol. 33: Asteraceae: Part 1: Lactuceae, Mutisieae, 'Tarchonantheae'
                       Part 2: Vernonieae, Cardueae
                       Part 3: Arctotideae
                       Part 4: Anthemideae
                       Part 5: Astereae
                       Part 6: Calenduleae
                       Part 7: Inuleae: Fascicle 1: Inulinae
                                          Fascicle 2: *Gnaphaliinae (First part)* (Published 1983). Price: 12,93. Other countries: R16,20
                       Part 8: Heliantheae, Eupatorieae
                       Part 9: Senecioneae

SIMAROUBACEAE

by K. L. IMMELMAN

4128 **KIRKIA**

Kirkia *Oliv.* in F.T.A. 1: 311 (1868); in Hooker's Icon. Pl. 11: t. 1036 (June, 1868); M. Friedrich in F.S.W.A. 69: 1 (1968); R. A. Dyer, Gen. 1: 295 (1975); Stannard in Kew Bull. 35,4: 829 (1981); Immelman in Bothalia 15, 1 & 2: 151 (1984). Type species: *K. acuminata* Oliv.

Trees; bark grey, with or without black spots, smooth, with salmon-pink lenticels. *Leaves* spiral, clustered at ends of branches, exstipulate, imparipinnate, deciduous. *Leaflets* opposite or subopposite, sessile or very shortly petiolulate, margins entire or crenate. *Inflorescence* an axillary dichasium. *Flowers* male or female, monoecious but flowers of only one sex open on a tree at a time. *Sepals* 4, free, ovate. *Petals* 4, free, oblong, adaxial surface glabrous or puberulous. *Stamens* 4, free, reduced in size and appear to be empty of pollen in female flowers. *Disc* 4-lobed, fleshy, saucer-shaped. *Stigma* capitate, slightly 4-lobed, absent in male flowers. *Styles* 4, fused, absent in male flowers, long in female flowers. *Ovary* 4-locular, each loculus with a single pendulous ovule, reduced in male flowers. *Fruit* oblong, 4-sided, dry, glabrous, separating into 4 or 8 one-seeded cocci each attached by a strip of tissue to top of central gynophore.

Kirkia is the only genus of the Simaroubaceae represented in the Flora area. It has 5 species, all African, 3 of which occur within the Flora area: in S.W.A./Namibia, Botswana and Transvaal.

The generic name commemorates Dr J. Kirk who collected the type of *K. acuminata*.

1a Leaves usually with fewer than 10 pairs of leaflets, leaflets 10 − 24 mm wide .... 1. *K. acuminata*
1b Leaves always with 10 or more pairs of leaflets, leaflets up to 6 mm wide:
  2a Bark with small grey-black spots; mericarps 8, fruit as long as wide; endemic to Sesfontein area in southern Kaokoland, S.W.A./Namibia .... 2. *K. dewinteri*
  2b Bark not spotted; mericarps 4, fruit 1½ times longer than wide; endemic to Transvaal...................... 3. *K. wilmsii*

1. **Kirkia acuminata** *Oliv.* in F.T.A. 1: 31 (1868); in Hooker's Icon. Pl. 11: t.1036 (June, 1868); Codd in Mem. bot. Surv. S. Afr. 26: 83, fig. 78, 79 (1951); Exell & Mendonça in C.F.A. 1: 276 (1956); Wild & Phipps in F.Z. 2: 214, t.39 (1963); De Winter *et al.*, Sixty-six Transv. Trees 86 (1966); M. Friedrich in F.S.W.A. 69: 2 (1968); Wild, Phipps & Paiva, F. M. 38 (Simaroubaceae): 3 (1969); Coates Palgrave, Trees of Southern Africa 345, t. 117 (1977). Type: Mozambique, from Batoka to the delta at Lupata near Sena, *Kirk* s.n. (not seen).

K. pubescens Burtt Davy, Fl. Transv. 1: 45 (1926); 2: 481 (1932). *K. acuminata* var. *pubescens* (Burtt Davy) Bremekamp in Ann. Transv. Mus. 15,2: 244 (1932−35). Type: Transvaal, Messina, *O'Connor* H 1934 (K, holo.; PRE!).

Tree with spreading rounded crown, 3−17 (−23) m high; bark pale grey, smooth or slightly rough. *Leaflets* in (3−) 6−9 (−10) pairs, glabrous or pilose, hairs sometimes restricted to midrib, elliptic, 29−88 × 10−24 mm, apex acuminate, slightly curving, base slightly oblique, margin crenate, becoming bright scarlet in autumn; petiolules absent to 2 mm long. *Pedicels* articulated near base, whitish-puberulous. *Sepals* glabrous or pubescent, 1−2 × 1−1,5 mm. *Petals* 4 (−5), glabrous or outer surface pubescent, 3−6 × 1−1,5 mm, white to cream. *Stamens* with anthers oval, 1−2 mm long; filaments linear, glabrous, 3−4 mm long; staminodes about ⅓ as large. *Fruit* glabrous, 11−20 × 6−9 mm, length 1,5−3 times the breadth. Fig. 1: 1.

A conspicuous and often common tree in bushveld and savanna, on rocky hillslopes; preferring basic soils (e.g. dolomite) but also found on a variety of acidic soils (e.g. granite, quartz, sandstone). It occurs in most of the Transvaal north of 25° S., and in northern S.W.A./Namibia, with a few scattered localities in northern Botswana. It flowers mainly in November and December. Map 1.

Vouchers: *De Winter* 2888 (WIND); *De Winter & Leistner* 5585; *Giess* 15005; *Mogg* 14481; *Pole Evans* 2590.

2. **Kirkia dewinteri** *Merxm. & Heine* in Mitt. bot. StSamml., Münch. 3: 617 (1960); Merxmüller in F.S.W.A. 2, 69 (1966−72); Coates Palgrave, Trees of Southern Africa 345 (1977). Types: S.W.A./Namibia, Kaokoveld, *Von Koenen* 104 (M, holo.; PRE!); Kaokoveld, 24,41 km from Warmbad on road to Ombombo, *De Winter & Leistner* 5837 (M; PRE!).

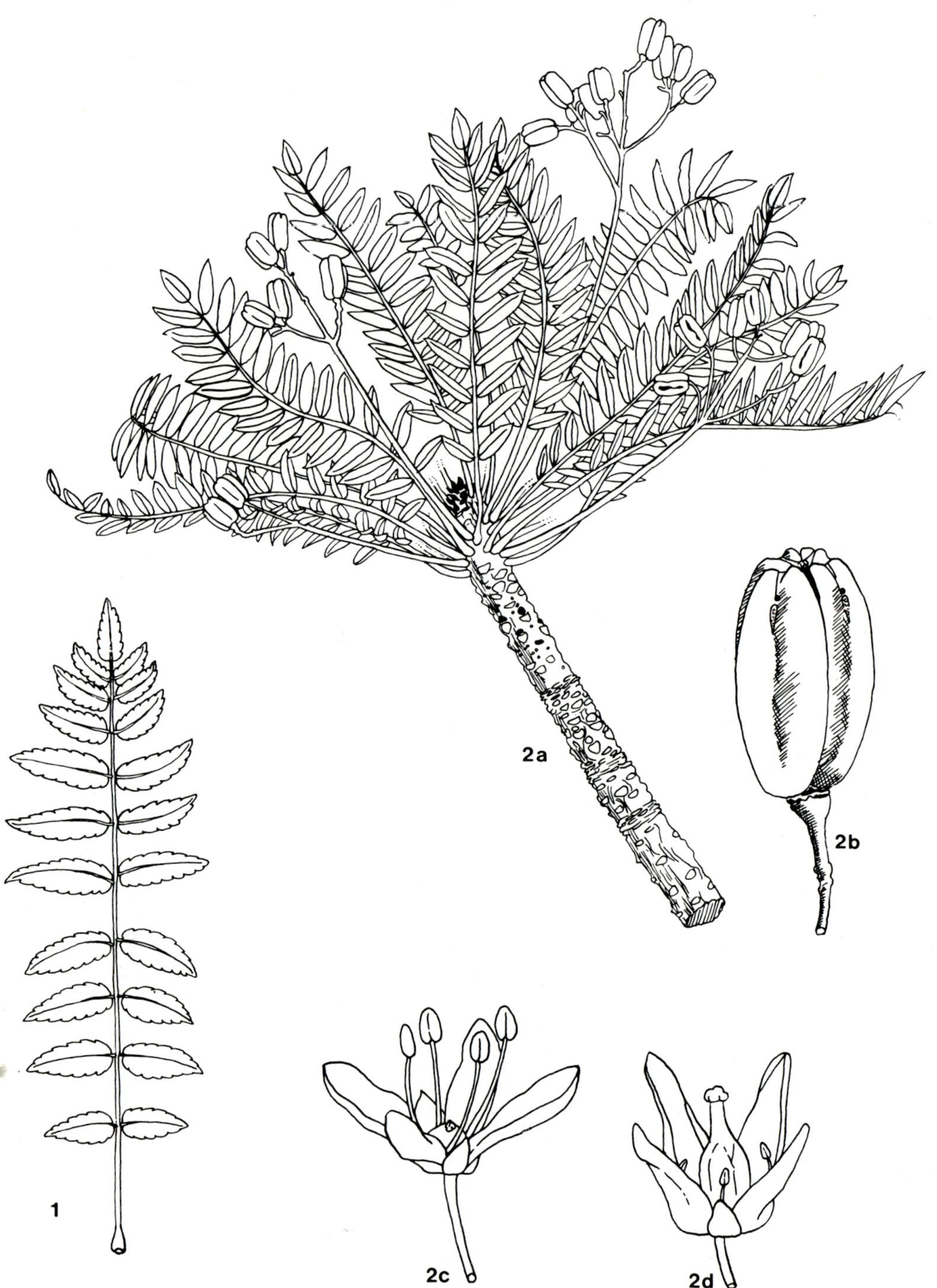

FIG. 1.—1, **Kirkia acuminata**: leaf, × 0,3 (*Story* 6514). 2, **Kirkia wilmsii**: 2a, fruiting branch, × 0,5 (*Codd* 794); 2b, fruit, × 2,7 (*Fourie* 2572); 2c, male flower, × 6,4 (*Fourie* 432); 2d, female flower, × 6,4 (*Van der Schijff* 1287).

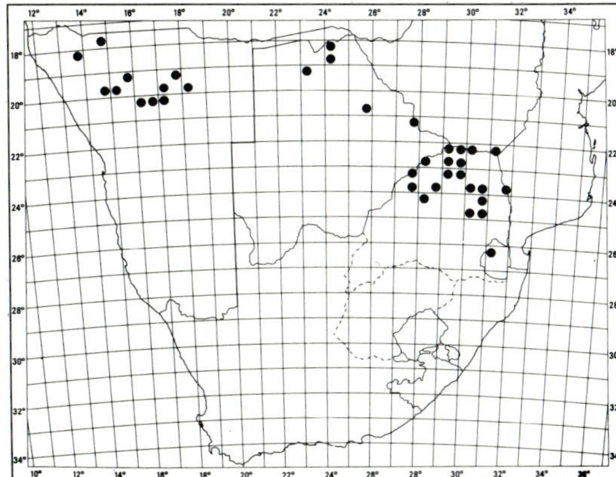

MAP 1.— **Kirkia acuminata**

Tree, 3—10 m high; bark smooth, yellowish with numerous small blackish spots. *Leaves* hysteranthous. *Leaflets* in 20—40 pairs, glabrous, elliptic, 6—16 × 2—3 mm, apex acute to acuminate, base slightly oblique, margin minutely crenate; petiolules absent or up to 1 mm long. *Flowers* coming out before the leaves, with pedicels articulated near base, whitish-puberulous. *Sepals* puberulous, 1,5—3 × 1—1,5 mm. *Petals* glabrous or puberulous on outer surface, 3,5—5 × 1,5—2 mm, cream. *Stamens* of male flowers with anthers oval, 1—1,5 mm long; filaments linear, glabrous, ±3 mm long; stamens of female flowers ½—⅓ as large. *Fruit* with a few hairs near the apex, c. 6 × 6 mm.

A rare tree confined to the Sesfontein area in the southern Kaokoland in S.W.A./Namibia. It occurs on the slopes of dolomitic hills and rock outcrops, and has been recorded in flower in September and December. Map 2.

Vouchers: *Craven* 1062 (WIND); *Giess* 10545; *Rusch* s.n. (WIND); *Van der Walt* 246 A.

3. **Kirkia wilmsii** *Engl.* in Notizbl. bot. Gart. Mus. Berl. 2: 25 (1897); Codd in Mem. bot. Surv. S. Afr. 26: 84, fig. 80 (1951); De Winter *et al.*, Sixty-six Transv. Trees 88

(1966); Ross in Flower. Pl. Afr. pl. 1590 (1970); Coates Palgrave, Trees of Southern Africa 345 (1977). Syntypes: *Wilms* 147, 148, 162 (B†).

Tree, crown rounded and widely spreading, trunk often branching near base, 3—20 m high; bark pale grey, smooth or slightly rough. *Leaflets* in 10—22 pairs, glabrous, elliptic, 6—30 × 2—6 mm, apex acute, base slightly oblique, margin usually minutely crenate, sometimes entire, becoming bright scarlet in autumn; petiolules absent. *Pedicels* articulated near base, usually glabrous. *Sepals* usually glabrous, 1—2 × 0,75—1 mm. *Petals* usually glabrous, 3—4 × 1—2 mm, cream to greenish white. *Stamens* of male flowers with anthers oval, 1—2 mm long; filaments linear, glabrous, 2—3 mm long; stamens of female flowers ½—⅓ as large. *Fruit* glabrous, 8—11 × 4—7 mm. Fig. 1: 2.

A common tree on rocky hillslopes and in kloofs, in savanna and bushveld, occurring in both granitic and dolomitic soils. It is endemic to the Transvaal Lowveld. It can be found in flower from September to January, but usually flowers from October to December. Map 2.

Vouchers: *De Winter* 8416; *Mogg* H. 11654; *Nel* 9 (NBG); *Robbertse* 1147; *Strey* 8001.

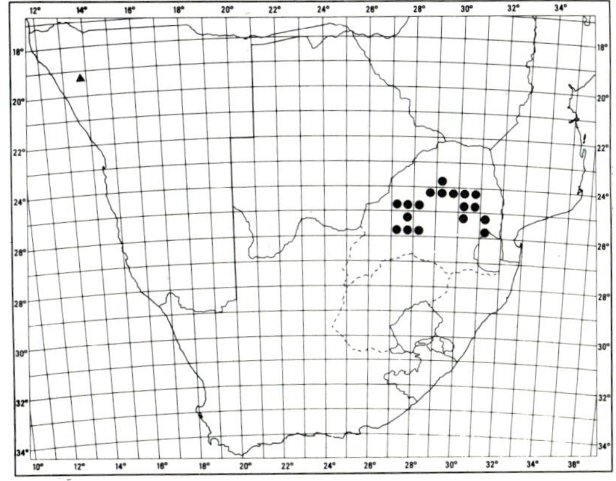

MAP 2.— ▲ **Kirkia dewinteri**
        ● **Kirkia wilmsii**

BURSERACEAE

by J. J. A. VAN DER WALT*

Dioecious or polygamous but rarely monoecious many-stemmed shrubs or trees; bark often peeling or flaking in papery pieces or strips; resin ducts present, secreting an odoriferous resin; branchlets often spine-tipped, glabrous, pilose or tomentose. *Leaves* petiolate or rarely sessile or subsessile, alternate, usually grouped at the ends of the branches, simple, trifoliolate or imparipinnate, margins of leaflets usually crenate, serrate or lobed, seldom entire. *Flowers* unisexual, rarely bisexual, perigynous or hypogynous; male flowers usually larger than female flowers; appearing before or with the leaves and occasionally after the leaves in axillary simple or compound dichasial cymes, paniculate cymes or singly in clusters. *Calyx* infundibuliform, campanulate or broadly campanulate with 4 valvate persistent lobes, in perigynous flowers continuous with hypanthium, in hypogynous flowers inserted on receptacle. *Petals* 4, usually yellow to green, apex incurved. *Disc* in perigynous flowers adnate to hypanthium, cylindrical, rarely fleshy, sometimes lobed; in hypogynous flowers not adnate to calyx or corolla, intrastaminal, cylindrical, usually with 4 large lobes or with 4 large and 4 small lobes. *Stamens* 8 or sometimes 4, obdiplostemonous, 4 antisepalous stamens longer than other 4; filaments inserted on the outside or on top of disc; anthers introrse and adnate; staminodes present in female flowers. *Gynoecium* half inferior or superior; rudimentary in male flowers; ovary 2-locular with 2 epitropous ovules per loculus; style usually relatively short; stigma capitate, obscurely 2−4-lobed. *Fruit* an ovoid, ellipsoid or subglobose drupe, usually asymmetrically flattened; exocarp relatively thin, mesocarp usually fleshy, exocarp and mesocarp splitting in ripe fruit into 2 longitudinal valves (4 valves in a few species outside our area); endocarp forming a crustaceous or bony putamen, usually clasped by a pseudo-aril; pseudo-aril usually red or yellowish, usually fleshy, cupular with short lobes or arms or with 2−4 relatively long arms or covering almost whole putamen without distinct arms; putamen usually enclosing one fertile loculus and a much smaller abortive loculus; seed with embryo straight, cotyledons much folded. Fig. 2.

Some characters not applicable in our area: sepals 3 or 5; petals 3 or 5, rarely absent; ovary 3–5-loculed; fruit sometimes a tardily dehiscing capsule.

A family of 16 genera with about 500 species distributed mainly in the tropics, but especially in Malaysia, Africa and South America.

4151   **COMMIPHORA**

Commiphora *Jacq.*, Hort. Schoenbr. 2: 66, t. 249 (1797); Engl. in A. DC., Monogr. Phan. 4: 7 (1883); in Pflanzenfam. 3: 251 (1896); edn 2, 19a: 429 (1931); Guillaumin in Annls Sci. nat. 9: 279 (1909); Engl. in Bot. Jb. 48: 449 (1913); in Pflanzenw. Afr. 3: 786 (1915); Hutch. & Dalz., F.W.T.A. 1: 488 (1928); Chiov., Fl. Somala 2: 53 (1932); Burtt in Kew Bull. 1935: 101 (1935); Perr. Bathie in Fl. Madag. 5: 5 (1946); Exell & Mendonça in C.F.A. 1: 298 (1951); Wild in Bolm Soc. broteriana, sér. 2, 33: 67 (1959); Wild in F.Z. 2: 263 (1963); Merxm. in F.S.W.A. 70: 1 (1968); J. J. A. v. d. Walt in Bothalia 11: 54 (1973). Type species: *C. madagascariensis* Jacq., Hort. Schoenbr. 2: 66, t. 249 (1797).

Amyris sensu Linn., Mant. 65 (1767).

Balsamea Gled. in Schriften Ges. naturf. Freunde, Berlin 3: 127 (1782); Engl. in Bot. Jb. 1: 41 (1881).

Balessan Bruce, Trav. 5: t. 25 (1790).

Balsamodendrum Kunth in Annls Sci. nat. 1: 348 (1824); DC., Prodr. 2: 76 (1825); Sond. in F.C. 1: 526 (1860); O. Berg in Bot. Ztg 21: 161 (1862); Marchand in Adansonia 8: 34, 67 (1867); Oliv. in F.T.A. 1: 324 (1868).

Hemprichia Ehrenb. in Linnaea 4: 396 (1829); Marchand in Adansonia 8: 69 (1867).

Heudelotia A. Rich. in Guill., Perr. & A. Rich., Fl. Sen. 1: 150, t. 39 (1832).

Protium sensu Wight & Arn. in Prod. Fl. Ind. 176 (1834); Harv. in F.C. 2: 592 (1862).

---

* Department of Botany, University of Stellenbosch.

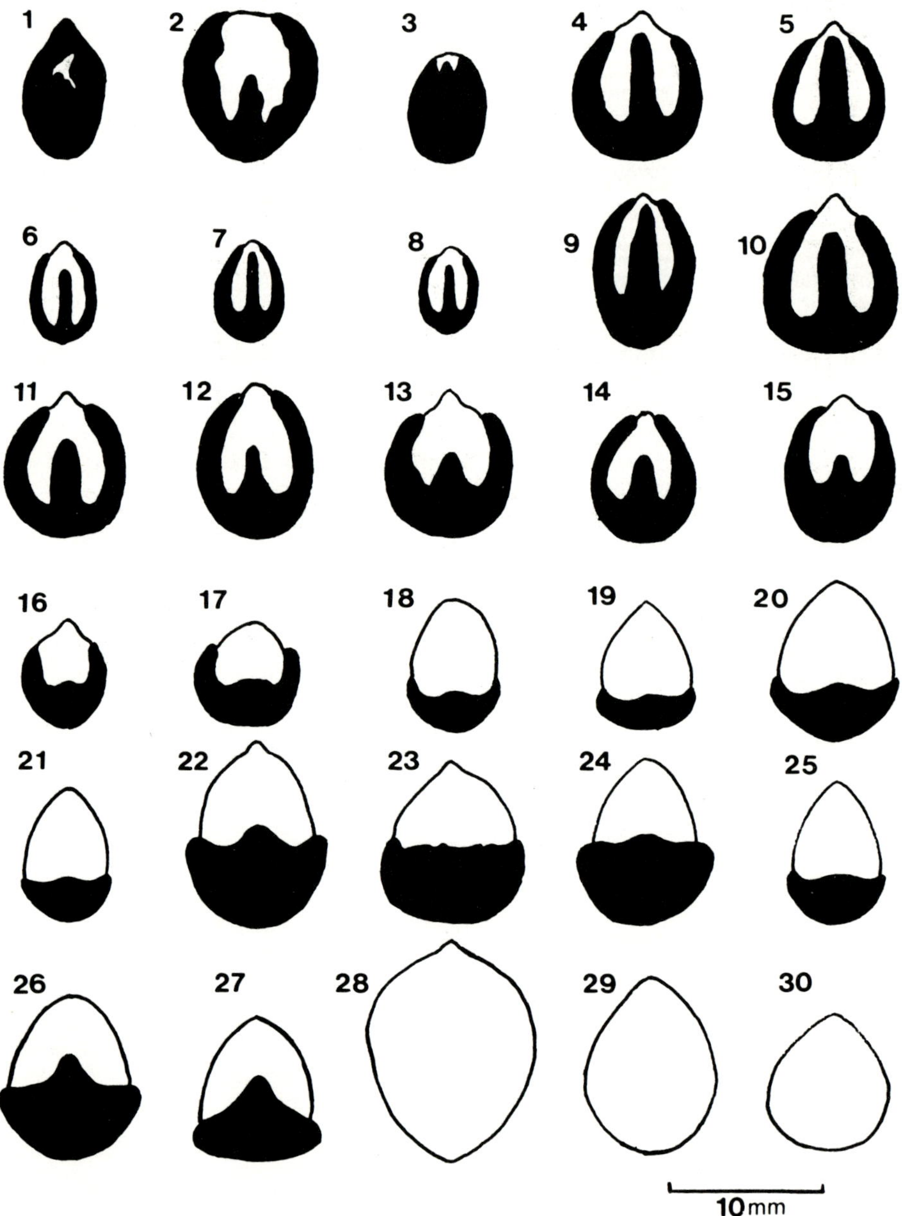

FIG. 2.—Putamens with pseudo-arils (in black) as seen from the less convex face of the putamens. 1, **Commiphora schimperi**; 2, **C. africana**; 3, **C. merkeri**; 4, **C. glandulosa**; 5, **C. pyracanthoides**; 6, **C. discolor**; 7, **C. virgata**; 8, **C. giessii**; 9, **C. multijuga**; 10, **C. neglecta**; 11, **C. mollis**; 12, **C. marlothii**; 13, **C. harveyi**; 14, **C. mossambicensis**; 15, **C. dinteri**; 16, **C. gracilifrondosa**; 17, **C. namaensis**; 18, **C. glaucescens**; 19, **C. wildii**; 20, **C. anacardiifolia**; 21, **C. saxicola**; 22, **C. edulis**; 23, **C. zanzibarica**; 24, **C. woodii**; 25, **C. crenato-serrata**; 26, **C. tenuipetiolata**; 27, **C. angolensis**; 28, **C. kraeuseliana**; 29, **C. capensis**; 30, **C. cervifolia**.

Protionopsin Blume, Mus. Bot. 1: 229 (1850), nom. nud.

Hitzeria Klotzsch in Peters, Reise Mossamb., Bot. 1: 89 (1861).

Balsamophloeos O. Berg in Bot. Ztg 20: 163 (1862).

Description as for family.

Commiphora, with about 200 species, is the only genus of the family represented in southern Africa.

1a Leaves simple, occasionally with 2 additional much smaller lateral leaflets on long shoots, leaves sessile or subsessile (except for *C. namaensis* (no. 18) with petiolate leaves):
  2a Branchlets not spine-tipped; flowers perigynous; pseudo-aril cupular without arms or with 2 commissural arms:
    3a Margin of leaves serrate-dentate, leaves petiolate, occasionally with 2 additional much smaller lateral leaflets, lamina rotund or orbicular and relatively small (up to 15×12 mm); pseudo-aril with 2 long commissural arms; bark light grey, not peeling.. 18. *C. namaensis*
    3b Margin of leaves entire, leaves sessile or subsessile, always simple, lamina narrowly elliptic to broadly elliptic and relatively large (usually much larger than 15×12 mm); pseudo-aril cupular with 3 or 4 short lobes but without 2 long commissural arms; bark yellowish brown or reddish brown and usually peeling in papery pieces:
      4a Leaves relatively large (70−)130(−200) × (30−)80(−140) mm; inflorescence thyrsoid, without large leaf-like bracts and up to 200 mm long; leaves and inflorescence pilose; fruit ovoid or obovoid....................
....21. *C. anacardiifolia*
      4b Leaves relatively small (15−)40(−100) × (8−)25(−60) mm; inflorescence of simple or compound dichasial cymes or thyrsoid, up to 80 mm long and often with large leaf-like bracts; leaves and inflorescence glabrous or sparsely to densely pilose; fruit ellipsoid............................... 19. *C. glaucescens*
  2b Branchlets spine-tipped; flowers hypogynous; pseudo-aril usually with 4 distinct arms of equal length reaching almost to apex of putamen:
    5a Young branchlets smooth and purplish; leaves glaucous and margins crenate-serrate; calyx glabrous; fruit ellipsoid and apiculate; bark grey ....3. *C. merkeri*
    5b Young branchlets greyish green; leaves green or dark green and margins finely crenate-serrate, crenate-serrate, subentire or entire; calyx with large glandular hairs or glabrous; fruit subglobose or ellipsoid but not apiculate; bark yellowish white, greyish green or yellowish green:
      6a Lamina of simple leaves broadly elliptic to suborbicular, dark green and shiny, margins often entire or subentire; calyx glabrous; flowers unisexual; bark yellowish white and peeling around the stem in papery strips .... 6. *C. discolor*
      6b Lamina of simple leaves elliptic, narrowly obovate or obovate, green and not shiny, margins usually finely crenate-serrate or serrate; calyx with large glandular hairs or glabrous; flowers bisexual or unisexual; bark greyish green or yellowish green and flaking in small papery pieces:
        7a Calyx glabrous; lamina of simple leaves/terminal leaflet elliptic, narrowly obovate or obovate, margin finely crenate-serrate or occasionally almost entire; many-stemmed shrubs, occasionally small trees with single trunk ....5. *C. pyracanthoides*
        7b Calyx with large glandular hairs; lamina of simple leaves or terminal leaflet obovate or elliptic, margin crenate-serrate or occasionally almost entire; trees with a single trunk, occasionally shrub-like
....4. *C. glandulosa*

1b Leaves trifoliolate or pinnate, petiolate:
  8a Branchlets spine-tipped; leaves trifoliolate:
    9a Branchlets and leaves pilose to tomentose; leaflets broadly obovate or elliptic; flowers unisexual; fruit subglobose; putamen rugose..2. *C. africana*
    9b Branchlets and leaves glabrous or with a few scattered short hairs; leaflets elliptic or ovate; flowers bisexual or unisexual; fruit subglobose or ellipsoid and apiculate; putamen rugose or smooth:
      10a Branchlets and leaves glabrous; leaflet margins coarsely crenate-serrate; flowers bisexual only; fruit ellipsoid and distinctly apiculate; pseudo-aril membranous and without distinct arms, putamen very rugose
.... 1. *C. schimperi*
      10b Branchlets and leaves with a few scattered short hairs; leaflet margins entire or distal half finely crenate-serrate; flowers bisexual or unisexual; fruit subglobose; pseudo-aril fleshy and with 4 arms, putamen smooth...10. *C. neglecta*
  8b Branchlets not spine-tipped; leaves trifoliolate or pinnate:
    11a Leaves and young branchlets hairy:
      12a Acroscopic margin of lateral leaflets incised to rachis but basiscopic margin decurrent along rachis 20. *C. wildii*
      12b Both margins of lateral leaflets incised to rachis or petiolule:
        13a Leaves pinnate, usually at least 6-jugate, leaflets abruptly acuminate at both ends, petiolules relatively long and slender; pseudo-aril cupular with 4 arms of equal length reaching almost to apex of putamen
....9. *C. multijuga*

13b Leaves trifoliolate or pinnate but then usually less than 6-jugate, leaflets not abruptly acuminate at both ends, petiolules of variable length; pseudo-aril cupular or with 2−4 arms or lobed:

14a Leaves pinnate, branchlets obtuse, petiole with medullary vascular bundles; fruit more than 15 mm in diameter:

15a Tree with a single main stem; bark peeling in large yellowish papery pieces; leaves dark green, leaflets obovate to broadly elliptic and not scabrous above; flowers hypogynous, disc pilose; pseudo-aril yellow with 2 long commissural and 2 short facial arms ....12. *C. marlothii*

15b Tree or many-stemmed shrub; bark flaking in small yellowish papery pieces; leaves greyish green, leaflets narrowly elliptic to narrowly ovate and scabrous above; flowers perigynous, disc glabrous; pseudo-aril red, cupular with 4 short lobes ....23. *C. edulis*

14b Leaves pinnate or trifoliolate; branchlets not obtuse, petiole without medullary vascular bundles; fruit usually less than 15 mm in diameter:

16a Margin of leaflets entire; flowers with a fleshy disc; pseudo-aril with 2 long commissural arms and 2 shorter facial arms; bark not papery:

17a Leaflets elliptic, oblong-elliptic or obovate, apex not abruptly acuminate; young branchlets, leaves and calyx without conspicuous golden glandular hairs; disc of flower 4-lobed .... 11. *C. mollis*

17b Leaflets ovate, broadly ovate, suborbicular or oblate, apex often abruptly acuminate; young branchlets, leaves and calyx with conspicuous golden glandular hairs; disc of flower 8-lobed .... 14. *C. mossambicensis*

16b Margin of leaflets usually at least partly crenate-serrate; flowers without a fleshy disc; pseudo-aril with 2 broad facial lobes; bark often papery:

18a Many-stemmed shrubs or bush; leaves always sparsely pilose to densely pubescent, petiole not slender; pseudo-aril covering lower ¼−½ of putamen....................................23. *C. angolensis*

18b Tree with a single trunk; leaves glabrous, sparsely pilose to pubescent, petiole often slender; pseudo-aril covering ⅓−¾ of putamen .... 27. *C. tenuipetiolata*

11b Leaves and young branchlets glabrous (irrespective of glandular hairs):

19a Leaves trifoliolate:

20a Leaflets linear, cultrate or narrowly oblanceolate and usually irregularly lobed, leaflets sessile or subsessile:

21a Leaves up to 80 mm long, leaflets linear to cultrate, margin coarsely dentate-serrate; branchlets slender; stamens 4 only; pseudo-aril present .... 16. *C. gracilifrondosa*

21b Leaves up to 20 mm long, leaflets cultrate or narrowly oblanceolate, margin entire (irrespective of lobes); branchlets short and stout; stamens 8; pseudo-aril absent.............................. 31. *C. cervifolia*

20b Leaflets of variable shape but never linear or cultrate and never irregularly lobed, leaflets petiolate, subsessile or sessile:

22a Margin of leaflets entire; flowers hypogynous with disc lobes not adnate to perianth; pseudo-aril with 4 arms of equal length reaching almost to apex of putamen; putamen with a prominent hump on the less convex face:

23a Shrub with many stems of *c.* 25 mm in diameter sprouting from ground level; bark reddish brown, usually not peeling...8. *C. giessii*

23b Shrub-like tree with a short trunk branching into relatively thick stems; bark yellowish white to silvery, peeling around the stem in papery strips ....7. *C. virgata*

22b Margin of leaflets not entire; flowers perigynous with disc lobes adnate to hypanthium; pseudo-aril absent or with arms or lobes of unequal length; putamen without a prominent hump on the less convex face:

24a Leaflets narrowly oblanceolate to oblanceolate; stamens 4 only.......................... 17. *C. oblanceolata*

24b Leaflets cordate, obovate, orbicular or elliptic; stamens 8:

25a Petiole usually relatively long and slender; pedicel longer than 3 mm; pseudo-aril with 2 facial lobes but without commissural arms; bark peeling .... 27. *C. tenuipetiolata*

25b Petiole relatively short and not slender; pedicel up to 1 mm long; pseudo-aril absent or with 2 commissural and 1−2 facial arms; bark usually not peeling:

26a Margin of leaflets undulate, crenate, or almost entire; pseudo-aril absent; fruit ellipsoid; shrub-like tree with a short trunk branching near ground level into thick stems...... 30. *C. capensis*

26b Margin of leaflets finely crenate-serrate; pseudo-aril with 2 commissural and 1−2 facial arms; fruit ovoid; shrubs with many relatively thin stems sprouting from ground level ....15. *C. dinteri*

19b Leaves pinnate (occasionally trifoliolate in *C. harveyi*—no. 13):

27a Leaflets linear; pseudo-aril absent .... 29. *C. kraeuseliana*

27b Leaflets not linear (wider); pseudo-aril present:

28a Leaflets asymmetrical and abruptly acuminate at both ends....................................9. *C. multijuga*

28b Leaflets symmetrical and not abruptly acuminate at both ends:
  29a Leaflets suborbicular to oblate ....22. *C. saxicola*
  29b Leaflets oblanceolate, lanceolate or elliptic to ovate:
    30a Leaflets with microscopic hairs; bark usually peeling in large brown papery pieces; flowers hypogynous, disc not adnate to calyx or corolla; pseudo-aril with 4 arms ....13. *C. harveyi*
    30b Leaflets glabrous (irrespective of glandular hairs); bark not peeling; flowers perigynous, disc adnate to hypanthium; pseudo-aril cupular without arms:
      31a Margin of leaflets entire (sometimes finely serrate) ....24. *C. zanzibarica*
      31b Margin of leaflets crenate-serrate to coarsely crenate-serrate (sometimes entire in *C. tenuipetiolata*—no. 27):
        32a Leaflets lanceolate or narrowly lanceolate; petioles with medullary vascular bundles; fruit ovoid and conspicuously beaked...26. *C. crenato-serrata*
        32b Leaflets narrowly elliptic to elliptic; petioles without medullary vascular bundles; fruit subglobose and not beaked:
          33a Young branches shallowly fluted and obtuse; bark not peeling; pedicel less than 1 mm long; pseudo-aril cupular with 1 very short facial lobe................................25. *C. woodii*
          33b Young branches not fluted and not obtuse; bark peeling; pedicel more than 1 mm long; pseudo-aril cupular with 2 facial lobes...27. *C. tenuipetiolata*

1. Commiphora schimperi *(O. Berg)*

Engl. in A. DC., Monogr. Phan. 4: 13 (1883); Schweinf. in Bull. Herb. Boissier 7,2: 288 (1899); Engl. in Pflanzenfam. edn 2,19a: 435, t. 204 C–D (1931); Wild in F.Z. 2: 277 (1963); J. J. A. v.d. Walt in Bothalia 11: 65, fig. 20–26 (1973). Syntypes: Ethiopia, Takazze, *Schimper* 624 (B†; K, lecto.!); Schoata, *Schimper* 1139 (B†; W!; G!); *Schimper* 1564 (B†; W!; G!).

Balsamodendrum schimperi O. Berg in Bot. Ztg 20: 162 (1862). *B. africanum* sensu Oliv., F.T.A. 1: 325 (1868), pro parte quoad specim. Schimper. *Balsamea schimperi* (O. Berg) Engl. in Bot. Jb. 1: 41 (1881).

Commiphora betschuanica Engl. in Bot. Jb. 44: 149 (1910); in Pflanzenfam. edn 2,19a: 435 (1931). Type: Botswana, Mugnune, *Seiner* 64 (B, holo.†; K, fragment!; BM, sketch!).

Shrub or small tree 2–6 m tall; bark peeling in black discs or flaking in small yellowish papery pieces; young branchlets glabrous, spine-tipped. *Leaves* trifoliolate, glabrous, green; petiole 5–30 mm long; leaflets elliptic to broad elliptic; petiolules up to 2 mm long; margins coarsely crenate-serrate especially in the upper half of leaflets, apex acute, base cuneate, terminal leaflet up to 50×35 mm, lateral leaflets up to 25×22 mm. *Inflorescence:* flowers borne in clusters. *Flowers* bisexual only, hypogynous. *Pedicel* 1–2 mm long and with a few glandular hairs. *Disc* 4-lobed, not adnate to perianth. *Stamens* 8. *Fruit* ellipsoid and distinctly apiculate, ± 17×10×10 mm, glabrous; putamen very rugose, pseudo-aril red, membranous, covering almost the whole putamen.

Occurs in Botswana, northern Tvl. and Zululand. It grows in savanna-woodland in well-drained, sandy soil. Also recorded from Zimbabwe, Mozambique, Tanzania, Kenya and Ethiopia. Map 3.

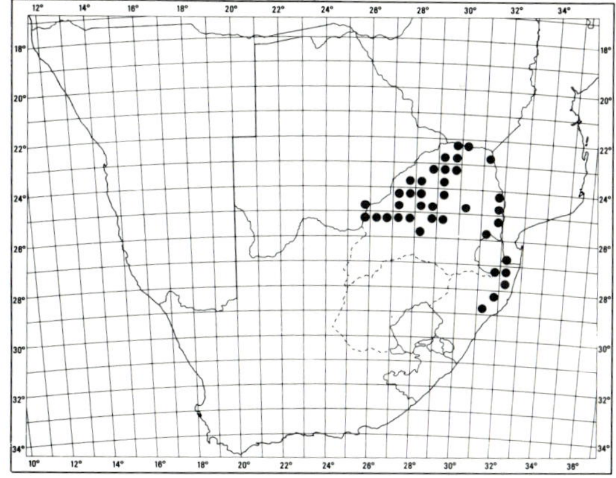

MAP 3.— **Commiphora schimperi**

Vouchers: *Acocks* 12993; *Codd* 6563; *De Winter & Killick* 8885; *Van der Walt* 11.

C. schimperi and *C. africana* (below), two closely related species, are often confused and this is mainly due to the similarity of the leaves. Besides many other differences such as in flower and fruit structure, the two species can also be distinguished on the hairiness of the leaves. The leaves of *C. africana* are pilose to tomentose, while those of *C. schimperi* possess only a few glandular hairs. A pungent resin odour is emitted when fresh leaves are picked.

Common name: Glossy-leaved Corkwood.

2. Commiphora africana *(A. Rich.)*

Engl. in A. DC., Monogr. Phan. 4: 14 (1883); Schweinf. in Bull. Herb. Boissier 7: 289 (1899); Hutch. & Dalz., F.W.T.A. 1: 488 (1928); Engl. in Pflanzenfam. edn 2,19a: 438 (1931); Exell & Mendonça in C.F.A. 1: 300 (1951); Irvine, Woody Plants Ghana 510 (1961); Wild in F.Z. 2: 276 (1963); Merxm. in

F.S.W.A. 70: 4 (1968); Lisowski, Malaisse & Symoens in Bull. Jard. bot. nat. Belg. 40: 357 (1970); J. J. A. v. d. Walt in Bothalia 11: 68, fig. 27–32 (1973); in Mitt. bot. StSamml., Münch. 12: 202, fig. 1 (1975). Type: Senegal, Kayar, *Leprieur* s.n. (P, holo., only photo seen; isotypes!).

Heudelotia africana A. Rich. in Guill., Perr. & A. Rich., Fl. Sen. 1: 150, t.39 (1832). *Balsamodendrum africanum* (A. Rich.) Arn. in Ann. nat. Hist. 3: 87 (1839); Oliv. in F.T.A. 1: 325 (1868), pro parte excl. syn. *B. schimperi* et vars.

Balsamodendrum kotschyi O. Berg in Bot. Ztg 20: 162 (1862). *Balsamea kotschyi* (O. Berg) Engl. in Bot. Jb. 1: 41 (1881). Type: Sudan, Nubia, *Kotschy* 271 (B, holo.†; K, lecto.!; S!; W!; MEL!; BM!).

Commiphora pilosa (Engl.) Engl. in A. DC., Monogr. Phan. 4: 12 (1883); in Pflanzenfam. edn 2,19a: 440 (1931); Chiov., Fl. Somala 2: 124, t.85 (1932); Palgrave, Trees Cent. Afr. 58, t. & photo (1956).

Balsamea pilosa Engl. in Bot. Jb. 1: 41 (1881). Type: Tanzania, Zanzibar, *Hildebrandt* 1184 (W, holo.!).

C. loandensis Engl. in Bot. Jb. 26: 370 (1899). Syntypes: Angola, Luanda, *Welwitsch* 4497 (LISU, lecto.!; BM!); 4498 (BM!; LISU!); 4498b (BM!; LISU!); 4500 (BM!; LISU!); 4501 (BM!; LISU!).

C. rubriflora Engl. in Bot. Jb. 30: 336 (1902). Type: Tanzania, near Rukwa Lake, *Goetze* 1406 (B, holo.†; K, fragment! and iso.!).

C. nkolola Engl. in Bot. Jb. 34: 308 (1905); in Pflanzenfam. edn 2,19a: 440 (1931). Type: Tanzania, Zanzibar coastal area, *Busse* 528 (B, holo.†; K, fragment!).

C. sambesiaca Engl. in Bot. Jb. 44: 146 (1910); in Pflanzenfam. edn 2,19a: 330 (1931); Burtt Davy, Fl. Transv. 2: 485 (1932). Type: Zambia, Kazungula, *Seiner* 90 (B, holo.†; K, fragment!).

C. calciicola Engl. in Bot. Jb. 44: 147 (1910); in Pflanzenfam. edn 2,19a: 440 (1931); Burtt Davy, Fl. Transv. 2: 485 (1932). Type: S.W.A./Namibia, Grootfontein; *Dinter* 820 (B, holo.†; K, fragment!; BM, sketch!).

Dioecious many-stemmed shrub usually less than 1 m tall or small tree with a single trunk up to 4 m tall; bark grey or greyish green, occasionally flaking locally in small yellowish papery pieces; young branchlets pilose to tomentose, mostly spine-tipped. *Leaves* trifoliolate, pilose to tomentose, green; petiole 2–35 mm long; leaflets broadly obovate or broadly elliptic, sessile or subsessile, margin coarsely crenate or crenate-serrate, apex obtuse to acute, base cuneate or truncate, terminal leaflet (8–)18(–65)×(6 –)13(–50) mm, lateral leaflets (4 –)8(–35)×(3–)7(–30) mm. *Inflorescence*: flowers borne in clusters. *Flowers* unisexual, hypogynous, glabrous. *Pedicel* 1–2,5 mm long. *Disc* 4-lobed, not adnate to perianth. *Stamens* 8. *Fruit* subglobose, ± 15×12×12 mm, glabrous; putamen very rugose; pseudo-aril red, with 4 arms of variable size and form, often

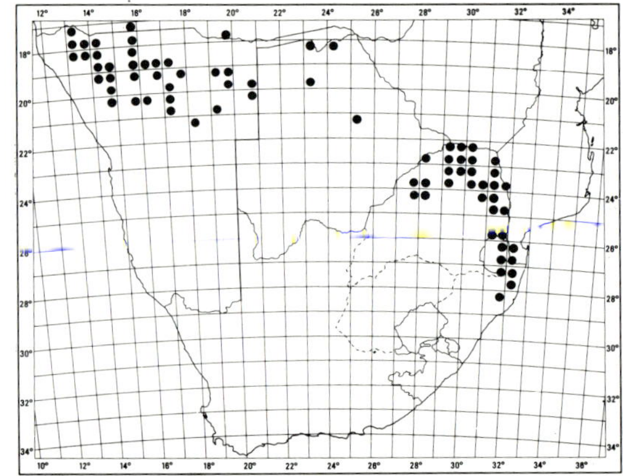

MAP 4.— **Commiphora africana**

also isolated fragments, 2 commissural arms reaching almost to apex of putamen, 2 facial arms of variable length.

Widely distributed in the northern part of S.W.A./Namibia, Botswana, northern Tvl., Swaziland and northern Zululand. Usually grows in sandy well-drained soil in shrub-thornveld, savanna-woodlands and broken mopaniveld.

Also recorded from Mozambique, Zimbabwe, Malawi, Zambia, Angola, Tanzania, Kenya, Ethiopia, Uganda, Sudan, Gambia, Senegal, Nigeria, Mauritania, Mali, Ghana, Togo, Niger, Zaire and Rwanda. Map 4.

Vouchers: *Compton* 29094; *De Winter* 3755; *Killick & De Winter* 8911; *Van der Schijff* 5237.

Wild (1963) distinguishes var. *africana* and var. *rubriflora* (Engl.) Wild. The calyx and pedicels of var. *rubriflora* are hairy, while those of var. *africana* are glabrous. As far as could be determined, only var. *africana* occurs in the F.S.A. area.

Wild (1963) mentions that the pseudo-aril of *C. africana* is apparently absent. However, all the fruits of the species studied possess a fleshy pseudo-aril.

According to Irvine (1961) the resin is used by the natives for perfuming and fumigating huts. He also mentions that it has several medicinal uses, and it is also used as a varnish. The species is easily grown from pole cuttings which are often planted as fencing poles.

Common name: Hairy Corkwood.

3. **Commiphora merkeri** *Engl.* in Bot. Jb. 44: 144 (1910); in Pflanzenfam. edn 2,19a: 437 (1931); Wild in F.Z. 2: 269 (1963); J. J. A. v.d. Walt in Bothalia 11: 63, fig. 14–19 (1973); in Mitt. bot. StSamml., Münch. 12: 205, fig. 3, 26a & a₁ (1975). Type: Tanzania, Nguruka, *Merker* 565 (B, holo.†; K, fragment!).

C. viminea Burtt Davy, Fl. Transv. 2: 485 (1932); Brenan in Kew Bull. 1953: 104 (1953). Type: Transvaal, Messina, *Moss & Rogers* 184b (K, holo.!).

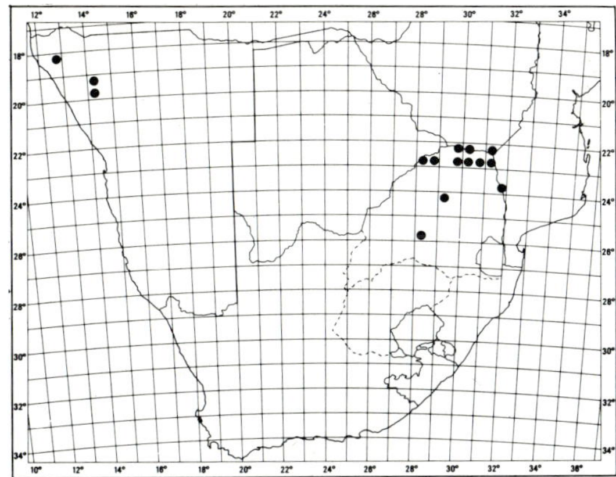

MAP 5.— **Commiphora merkeri**

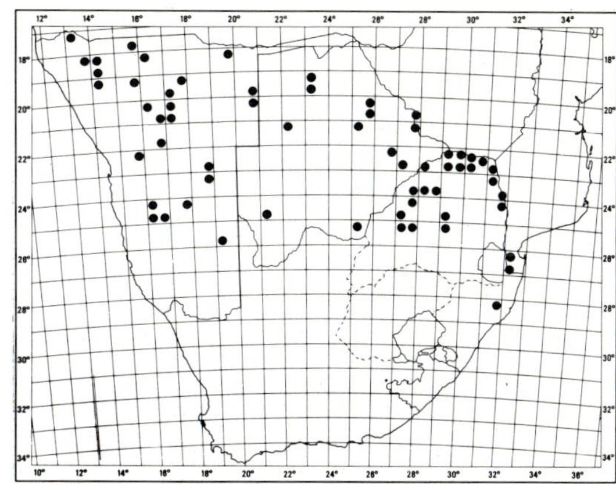

MAP 6.— **Commiphora glandulosa**

Dioecious tree 2−5 m tall with a single trunk; bark grey with dark patches, peeling around the stem in yellowish papery strips; young branchlets glabrous, smooth, purplish, often spine-tipped. *Leaves* simple but on long shoots often trifoliolate with smaller lateral leaflets, with long glandular hairs at base but otherwise glabrous, glaucous, subsessile, margin crenate-serrate especially near apex, apex acute to obtuse, base cuneate, lamina of simple leaves/terminal leaflet narrowly obovate to obovate or elliptic, 7−45×5−25 mm, lateral leaflets elliptic, 5−8×3−5 mm. *Inflorescence*: flowers borne in clusters. *Flowers* unisexual, hypogynous, glabrous. *Pedicel* 2−10 mm long. *Disc* 4-lobed, not adnate to perianth. *Stamens* 8. *Fruit* ellipsoid, 10×6×5 mm, glabrous, apiculate; putamen rugose; pseudo-aril yellow, covering the whole putamen except the apex.

Occurs in northern Tvl. from the border of Botswana in the west to Mozambique in the east, but it is particularly common north of the Soutpansberg. It is known from a few localities in Kaokoland, S.W.A./Namibia. Also recorded from Zimbabwe, Mozambique, Tanzania and Kenya. Map 5.

Vouchers: *Codd* 4111, 4835; *Giess* 7727; *Van der Walt* 6.

C. merkeri could be conspecific with *C. ellenbeckii* Engl. and probably also with a few other central African species. The fragment of the type specimen of *C. merkeri* at Kew is very poor, and little is known of the pseudo-arils of these allied species.

The stems often exude large quantities of resin.

Common name: Zebra-bark Corkwood.

4. **Commiphora glandulosa** *Schinz* in Bull. Herb. Boissier, sér. 2,8: 633 (1908); Exell & Mendonça in C.F.A. 1: 298 (1951); J. J. A. v.d. Walt in Bothalia 11: 57, fig. 1−7 (1973); in Mitt. bot. StSamml., Münch. 12: 206, fig. 4 (1975). Syntypes: S.W.A./Namibia, Ombandja, *Schinz* 767 (Z, lecto.!); Ondonga, *Schinz* s.n. (not seen).

C. pyracanthoides Engl. subsp. *glandulosa* (Schinz) Wild in Bolm Soc. broteriana, sér. 2, 33: 44 (1959); in F.Z. 2: 268 (1963); Von Breitenbach, Ind. Trees S. Afr 3: 433 (1965); Merxm. in F.S.W.A. 70: 8 (1968).

C. lugardae N.E. Br. in Kew Bull. 1909: 99 (1909); Miller in J1 S. Afr. Bot. 18: 38 (1952). Type: Botswana, Kwebe Hills, *Lugard* 23 (K, holo.!).

C. seineri Engl. in Bot. Jb. 44: 145 (1910); in Bot. Jb. 48: 480 (1913); in Pflanzenfam. edn 2,19a: 437 (1931). Type: Zambia, Sesheko, *Seiner* 57 (B, holo.!; K, photo!; BM, sketch!).

C. berberidifolia Engl. in Bot. Jb. 48: 480 (1913); in Pflanzenfam. edn 2,19a: 437 (1931). Type: S.W.A./Namibia, Okahandja, Waldau, *Dinter* 385 (B, holo.!; K, fragment!).

Polygamous or dioecious tree with a single trunk, 2−10 m tall, occasionally shrub-like; bark yellowish green or greyish green, flaking in small yellowish papery pieces; young branchlets glabrous, spine-tipped. *Leaves* usually simple but on long shoots often trifoliolate with smaller lateral leaflets, with long glandular hairs at base but otherwise glabrous, green, subsessile, margin crenate-serrate, occasionally almost entire, apex acute or obtuse, base cuneate, lamina of simple leaves/terminal leaflet obovate or elliptic (20−)35(−65)× (12−)25(−45) mm, lateral leaflets elliptic, (8−)15(−30)×(4−)7(−15) mm. *Inflorescence*: reduced cymes or flowers borne in clusters. *Flowers* bisexual, occasionally unisexual, hypogynous. *Pedicel* 0,5−1 mm long, pedicel and calyx with large glandular hairs. *Disc* 4-lobed, not folded, inside of lobes not grooved,

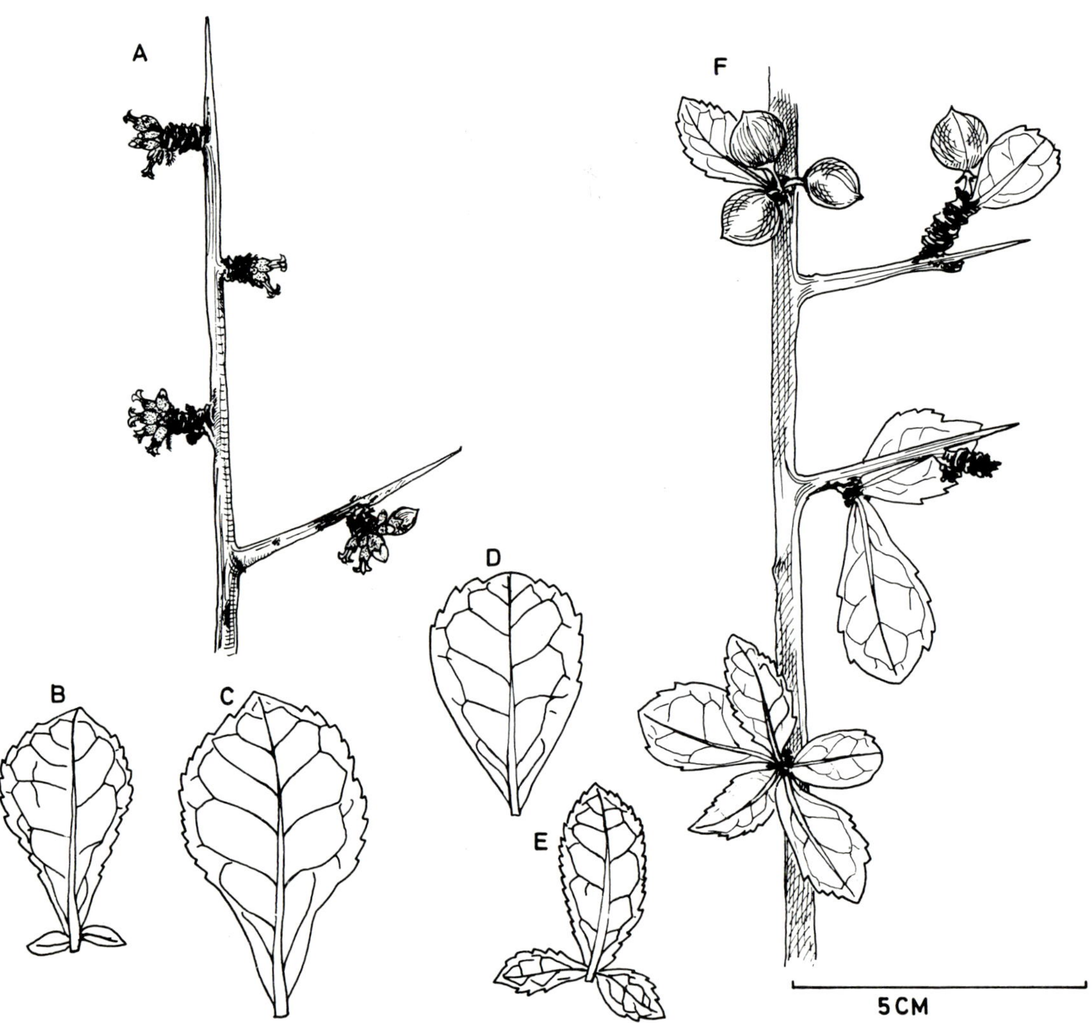

FIG. 3.—**Commiphora glandulosa:** A, branchlet with flowers and young fruits; B–E, leaves; F, branchlet with leaves and mature fruits.

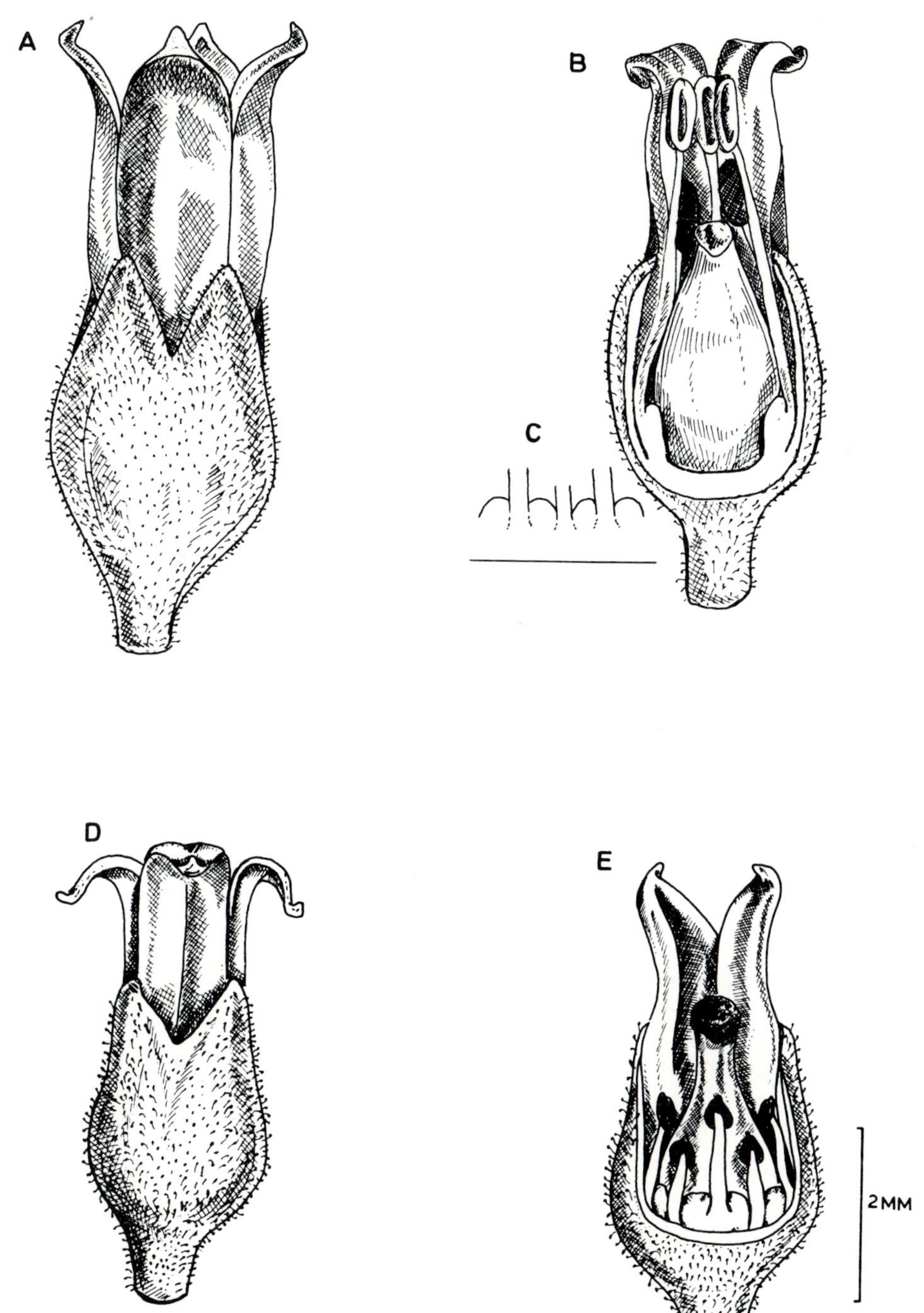

FIG. 4.—Flowers of **Commiphora glandulosa:** A, bisexual flower; B, bisexual flower with calyx and corolla partly removed; C, diagrammatic representation of two disc lobes illustrating the insertion of the filaments; D, female flower; E, female flower with the calyx and corolla partly removed.

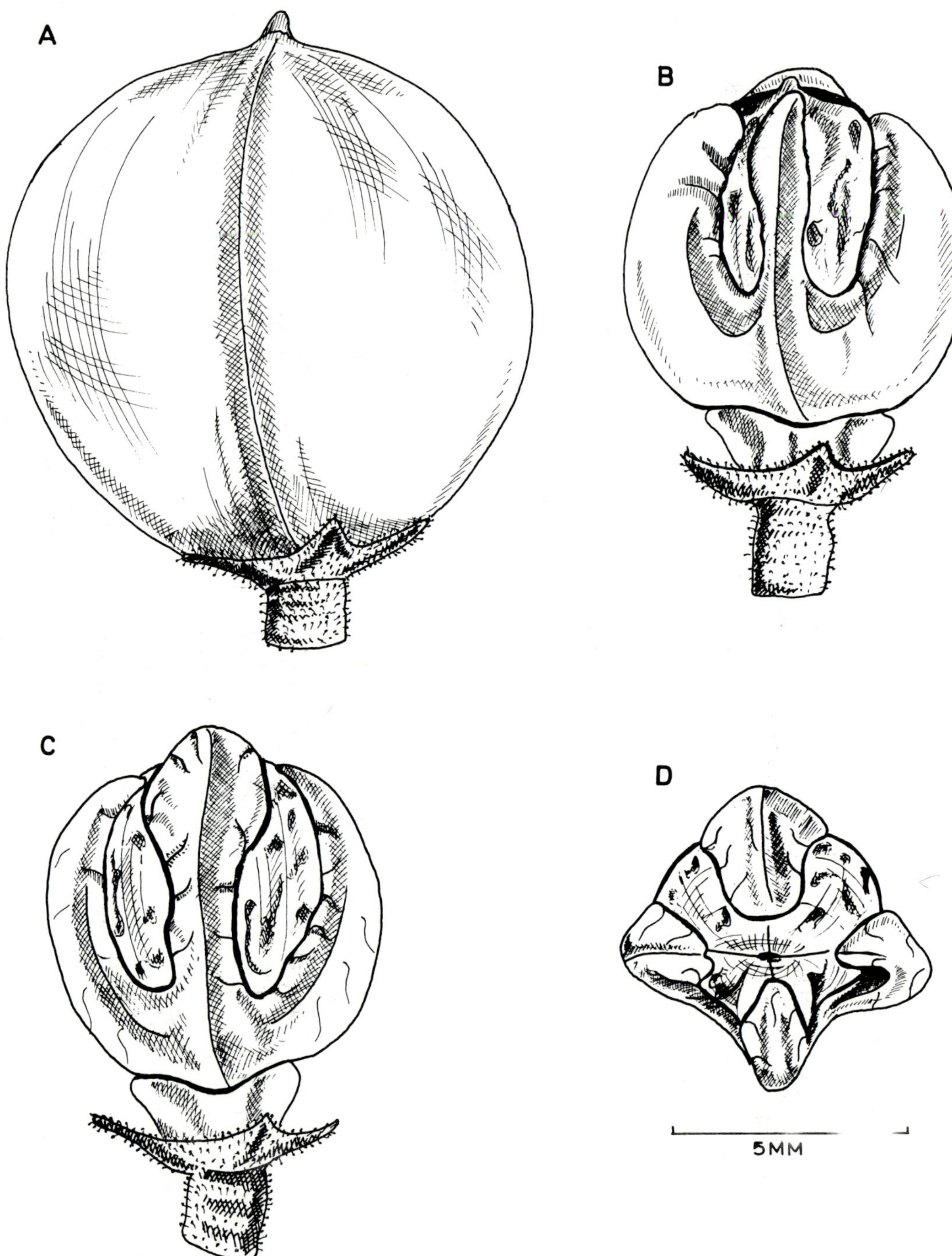

FIG. 5.—Fruit of **Commiphora glandulosa:** A, side view of the fruit; B, view of the less convex face of putamen with pseudo-aril; C, view of the more convex face of putamen with pseudo-aril; D, putamen and pseudo-aril as seen from above.

not adnate to perianth. *Stamens* 8. *Fruit* subglobose, ± 14×13×12 mm, glabrous; putamen rugose, with a hump on less convex face; pseudo-aril red, with 4 arms of equal length reaching almost to apex of putamen. Figs 3, 4 & 5.

Widely distributed in S.W.A./Namibia, Botswana, northern Tvl. and northern Zululand. Particularly common north of the Soutpansberg. Grows in sandy, well-drained soil in savanna-woodland or in broken mopaniveld. Also recorded from Zimbabwe, Zambia, Mozambique and Angola. Map 6.

Vouchers: *Codd* 4680; *De Winter* 4041; *Merxmüller & Giess* 830; *Van der Walt* 14.

Wild in Bolm Soc. broteriana, sér. 2, 33: 43; 1959 considers this taxon to be a subspecies of *C. pyracanthoides* Engl. (no. 5). This taxonomic change by Wild is based mainly on observations made by Merxmüller in S.W.A./Namibia where *C. glandulosa* occurs in tree and shrub form. However, the flower, and fruit structure of these two taxa differ to such a degree that they are considered to be different species.

It is easily grown from pole cuttings which are often planted as fencing poles.

Common name: Tall Common Corkwood.

5. **Commiphora pyracanthoides** *Engl.* in Bot. Jb. 26: 368 (1899); in Pflanzenfam. edn 2,19a: 437 (1931); Burtt Davy, Fl. Transv. 2: 485 (1932); Wild in Bolm Soc. broteriana, sér. 2, 33: 43, 82(1959); Wild in F.Z. 2: 268 (1963); Merxm. in F.S.W.A. 70: 8 (1968); J. J. A. v.d. Walt in Bothalia 11: 60, fig. 8−13 (1973); in Mitt. bot. StSamml., Münch. 12: 208, fig. 5, 29a & a₁ (1975). Type: S.W.A./Namibia, Otjimbingwe, *Fischer* 8 (holo.†; ?); Neotype: S.W.A./Namibia, Little Karas Mountains, Holoog, *Pearson* 9747 (K!).

Dioecious or polygamous many-stemmed shrub, 0,5−3 m tall, occasionally a small tree with single trunk up to 3 m tall; bark greyish green or yellowish green, flaking in small yellowish papery pieces; young branchlets glabrous, spine-tipped. *Leaves* usually simple but on long shoots often trifoliolate with smaller lateral leaflets, with long glandular hairs at base but otherwise glabrous, green, subsessile, margin finely crenate-serrate, occasionally almost entire, apex acute or obtuse, base cuneate, lamina of simple leaves/terminal leaflet elliptic, narrowly obovate or obovate, (16−)25(−55)× (8−)16(−32) mm, lateral leaflets narrowly elliptic or elliptic, (4−)8(−12)×(2−)3(−10) mm. *Inflorescence*: reduced cymes of flowers borne in clusters. *Flowers* unisexual or bisexual, hypogynous. *Pedicel* 0,5−1 mm long, pe-

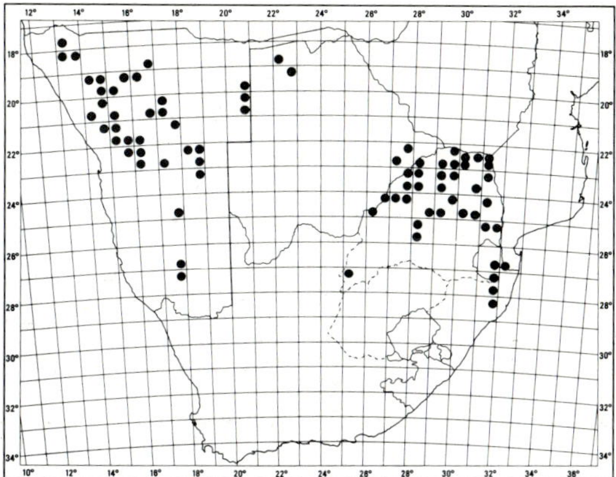

MAP 7.— **Commiphora pyracanthoides**

dicel and calyx without glandular hairs (glabrous). *Disc* 4-lobed, folded to form 4 large lobes towards the outside, inside of lobes deeply grooved, not adnate to perianth. *Stamens* 8. *Fruit* ellipsoid to subglobose, ± 11×8×7 mm, glabrous; putamen rugose, with a hump on less convex face; pseudo-aril red, with 4 arms of equal length reaching almost to apex of putamen.

Widely distributed in S.W.A./Namibia, Botswana, northern Tvl., Swaziland and northern Zululand. Known from a few localities in northern Cape. It grows in sandy, well-drained soil in savanna-woodland, shrub-thornveld and broken mopaniveld. Also recorded from Zimbabwe and Mozambique. Map 7.

Vouchers: *Codd* 4431; *De Winter & Leistner* 5115; *Van der Schijff* 5217; *Van der Walt* 57, 76.

In certain areas of S.W.A./Namibia it is impossible to distinguish on habit alone between *C. glandulosa* (no. 4) and *C. pyracanthoides* because both species can be either shrub-like or small trees. When occurring together in the same area, however, they can usually easily be distinguished as *C. glandulosa* is a small tree with a single bole and *C. pyracanthoides* a small, often many-stemmed shrub.

The most reliable character for distinguishing flowering material of the two species, are the glandular hairs which occur on the calyx of only *C. glandulosa*. Flowers should be examined carefully as the number of glandular hairs varies considerably, and they are also usually present on the bracteoles of both species.

Common name: Common Corkwood.

6. **Commiphora discolor** *Mendes* in Bolm Soc. broteriana, sér. 2, 41: 155; t. 1 & 2 (1967); Merxm. in F.S.W.A. 70: 6 (1968); J. J. A. v.d. Walt in Madoqua ser. 1: 10, t. 11−13 (1974); in Mitt. bot. StSamml., Münch. 12: 204, fig. 2 (1975). Type: Angola, Huila, *Mendes* 1693 (LISC, holo.!; BM!; COI!).

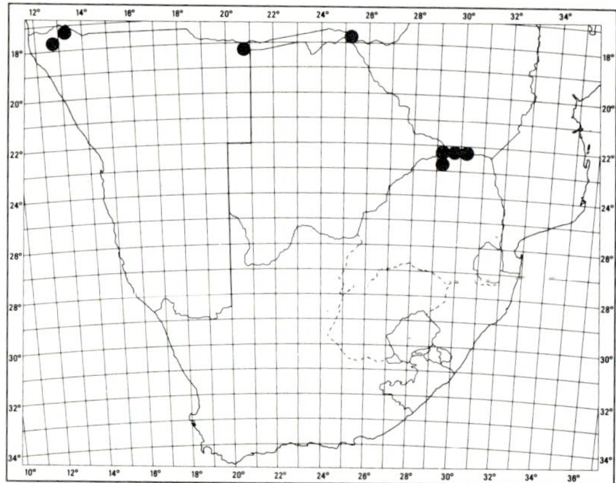

MAP 8.— **Commiphora discolor**

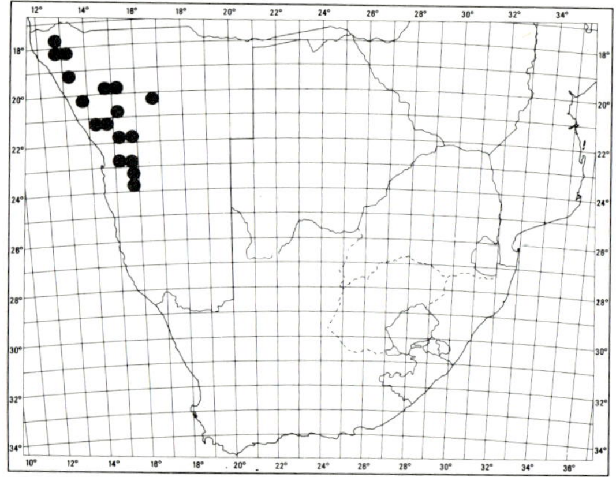

MAP 9.— **Commiphora virgata**

Dioecious tree 3−9 m tall with a single trunk; bark yellowish white, peeling around the stem in papery strips; young branchlets glabrous, often spine-tipped. *Leaves* usually simple but on long shoots often trifoliolate with smaller lateral leaflets, glabrous, dark green and shiny, subsessile, margin entire, subentire or crenate-serrate, lamina of simple leaves broadly elliptic to suborbicular, 30−60× 25−40 mm, apex truncate or acute, base truncate or cuneate, leaflets of trifoliolate leaves elliptic to broadly elliptic, sessile or subsessile, apex acute, base cuneate, terminal leaflet 28−70×12−50 mm, lateral leaflets 14−32× 7−15 mm. *Inflorescence*: flowers borne in reduced cymes, glabrous. *Flowers* unisexual, hypogynous, subsessile, glabrous. *Disc* not seen. *Stamens* 8. *Fruit* subglobose, ± 9×8×8 mm, glabrous; putamen smooth; pseudo-aril with 4 (3) arms of equal length reaching almost to apex of putamen.

Apparently confined to the north-western part of Kaokoland in S.W.A./Namibia and Angola. It grows on rocky hills and mountain slopes. Map 8.

Vouchers: *De Winter & Leistner* 5490; *Van der Walt & Giess* 291.

The papery bark peeling around the stem resembles that of *C. merkeri* (no. 3). Mendes (1967) mentioned instances of a climbing habit in his description of this species.

Common name: Kaoko Corkwood.

7. **Commiphora virgata** *Engl.* in Bot. Jb. 19: 139 (1894); in Pflanzenfam. edn 2, 19a: 433 (1931); Exell & Mendonça in C.F.A. 1: 300 (1951); Merxm. in F.S.W.A. 70: 9 (1968); J. J. A. v.d. Walt in Dinteria 9: 23, 24 (1973); in Madoqua ser. 1: 20, t. 34−36 (1974); in Mitt.

bot. StSamml., Münch. 12: 209, fig. 6, 26b & b₁ (1975). Type: S.W.A./Namibia, Sorris-Sorris, *Gürich* 68 (B, holo. †; BM, sketch!).

Dioecious shrub-like tree 0,5−3 m tall, branching near ground level into relatively thick stems; bark yellowish white to silvery, peeling around the stems in papery strips of the same colour; young branchlets glabrous, slender and often drooping, not spine-tipped. *Leaves* trifoliolate, glabrous; petiole 2−15 mm long; leaflets narrowly obovate, elliptic or broadly elliptic; petiolules less than 1 mm long; margin entire, apex obtuse, seldom retuse or acute, base cuneate, terminal leaflet (4−)12(−25)×(2−)7 (−10) mm, lateral leaflets (4−)10(−15)×(2−) 5(−7) mm. *Inflorescence*: reduced cymes, glabrous, or flowers solitary. *Flowers* unisexual, hypogynous, subsessile, glabrous. *Disc* 4-lobed, not adnate to perianth. *Stamens* 8. *Fruit* irregularly subglobose, ellipsoid or obovoid, ± 8×7×7 mm, glabrous; putamen rugose, with a prominent hump on upper half of less convex face; pseudo-aril white to reddish, with 4 arms of equal length reaching almost to apex of putamen, in some cases also isolated fragments.

Occurs in S.W.A./Namibia on the edge of the Namib Desert, and it has been collected from Ombepera in the north to Solitaire in the south. It usually grows on rocky hills or stony slopes. Map 9.

Vouchers: *De Winter* 3147; *De Winter & Leistner* 5663; *Merxmüller & Giess* 928.

C. virgata is closely related to *C. giessii* (no. 8) and their leaves, flowers and fruits are very similar. The differences between the two species have been discussed in detail by Van der Walt (1973).

Common name: Slender Corkwood.

8. Commiphora giessii *J. J. A. v.d. Walt* in Dinteria 9: 23−28, fig. 1−5 (1973); in Madoqua ser. 1: 11, t. 14−16 (1974); in Mitt. bot. StSamml., Münch. 12: 210, fig. 7 (1975). Type: S.W.A./Namibia, near Sesfontein, *Van der Walt* 242 (PRE, holo.; WIND; STE; M).

Dioecious shrub, 1,5−3 m tall, many stems of ± 25 mm in diameter sprouting from ground level; bark reddish brown and usually not peeling; young branchlets glabrous, very slender, often drooping, not spine-tipped. *Leaves* trifoliolate, glabrous; petiole 5−25 mm long; leaflets elliptic to narrowly obovate; petiolules less than 1 mm long; margin entire, apex acute or obtuse, base cuneate, terminal leaflet (10−)12(−45)×(5−)10(−25) mm, lateral leaflets (10−)15(−35)×(5−)7(−15) mm. *Inflorescence*: reduced cymes, glabrous, or flowers solitary. *Flowers* unisexual, hypogynous, subsessile, glabrous. *Disc* 4-lobed, not adnate to perianth. *Stamens* 8. *Fruit* irregularly obovoid or subglobose, ± 6×5×5 mm, glabrous; putamen slightly rugose, with a prominent hump on upper half of less convex face; pseudo-aril reddish, with 4 thin arms of equal length reaching almost to apex of putamen.

Confined to Kaokoland in S.W.A./Namibia where it is known from Okonjombo to Sesfontein. It is very common north-west of Sesfontein where it grows on the slopes of mountains, on hills and also in valleys. This area is hot and arid with an annual rainfall of ± 250 mm. Map 10.

Vouchers: *De Winter & Leistner* 5713, 5869; *Giess & Leippert* 7418.

C. giessii differs from *C. virgata* (no. 7) in habit and bark characteristics. The short trunk of *C. virgata* branches into relatively thick stems, and its bark peels off in yellowish white to silvery papery strips (Van der Walt, 1973).

Common name: Brown-stem Corkwood.

9. Commiphora multijuga *(Hiern) K. Schum.* in Just's bot. Jber. 27: 480 (1901); Exell & Mendonça in C.F.A. 1: 302 (1951); Merxm. in F.S.W.A. 70: 7 (1968); J. J. A. v.d. Walt in Madoqua ser. 1: 15, t. 24−26 (1974); in Mitt. bot. StSamml., Münch. 12: 213, fig. 27a & a_1, 29C & C_1 (1975). Type: Angola, Mossamedes, between Cazimba and Pomangala, *Welwitsch* 4503 (BM, holo.!; LISU!).

Dioecious tree 3−8 m tall, usually with a single trunk; bark reddish grey to dark grey, smooth, not peeling but in some cases cracked on old trunks; young branchlets sparsely pubescent or pubescent. *Leaves* pinnate, 4−10-jugate, sparsely pubescent or almost glabrous; petiole 15−40 mm long, slender; leaflets asymmetrically elliptic, broadly elliptic or rotund but abruptly acuminate at both ends, (12−)18(−25) × (10−)13(−20) mm; petiolules 5−15 mm long, slender, margin entire, apex acute, base cuneate. *Inflorescence*: simple or compound dichasial cymes or thyrsoid, up to 50 mm long, glabrous or sparsely pubescent. *Flowers* unisexual, perigynous. *Pedicel* 1−4 mm long, pedicel and calyx glabrous or sparsely pubescent. *Petals* conspicuously recurved. *Disc* 4-lobed, upper part of lobes free but lower part adnate to hypanthium. *Stamens* 8. *Fruit* subglobose, c. 15 × 15 × 14 mm, glabrous; putamen smooth; pseudo-aril red, cupular with 4 arms of equal length reaching almost to apex of putamen.

Occurs in S.W.A./Namibia, mainly in Kaokoland from the Kunene River southwards to Welwitschia. Also recorded from Angola. Map 11.

Vouchers: *De Winter & Leistner* 5233; *Merxmüller & Giess* 1383; *Van der Walt & Giess* 297.

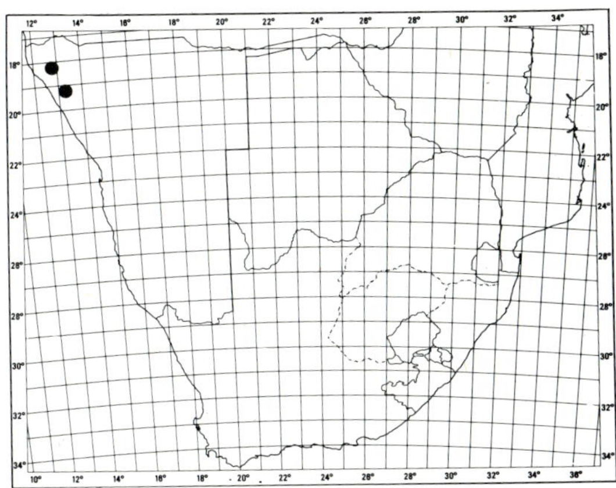

MAP 10.— **Commiphora giessii**

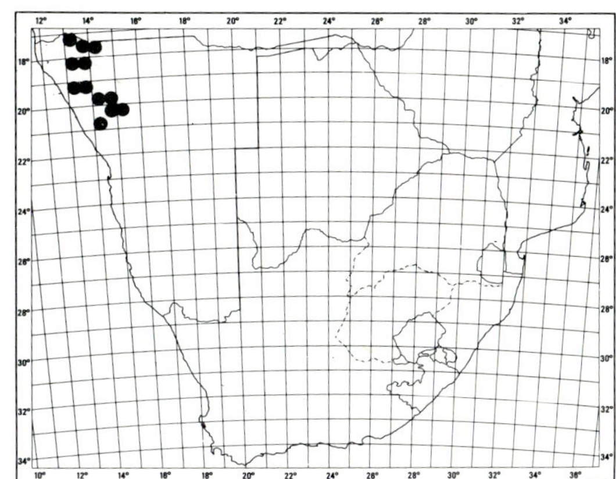

MAP 11.— **Commiphora multijuga**

C. multijuga is distinguished by its typical pale green-ish leaves which contrast well against the dark bark. The graceful leaves with drooping leaflets are very aromatic. Large quantities of colourless, pungent and sticky resin exude when twigs are cut.

Common name: Purple-stem Corkwood.

10. **Commiphora neglecta** *Verdoorn* in Bothalia 6: 214 (1951); Wild in F.Z. 2: 271 (1963); J. J. A. v.d. Walt in Bothalia 11: 71, fig. 33—38 (1973). Type: Transvaal, Skukuza, *Codd & Verdoorn* 5498 (PRE, holo.!).

Polygamous or dioecious many-stemmed shrub or small tree up to 8 m tall; bark grey to green, smooth or flaking in small yellowish papery pieces; young branchlets with a few short hairs, spine-tipped. *Leaves* trifoliolate, with a few short hairs, green; petiole 5—45 mm long; leaflets elliptic or ovate to broadly ovate; petiolules usually less than 1 mm long; margins entire or upper half finely crenate-serrate, apex acute, base cuneate, terminal leaflet up to 45 × 30 mm, lateral leaflets up to 30 × 22 mm. *Inflorescence*: axillary dichasial cymes or flow-ers borne in clusters. *Flowers* bisexual or uni-sexual, hypogynous, *Pedicel* 2—5 mm long, pedicel and calyx often with a few short hairs. *Disc* 4-lobed, not adnate to perianth. *Stamens* 8. *Fruit* subglobose, ± 15 × 14 × 14 mm, glabrous; putamen smooth; pseudo-aril red, with 4 arms, 2 commissural arms reaching al-most to apex of putamen, 2 facial arms shorter.

Occurs in central and northern Tvl., and is widely distributed in Natal, being particularly common in northern Zululand. It usually grows on the slopes of mountains or in sandy, well-drained soil. Also recorded from Mozambique. Map 12.

Vouchers: *Codd* 4821, 6641; *De Winter* 8985; *Van der Walt* 69.

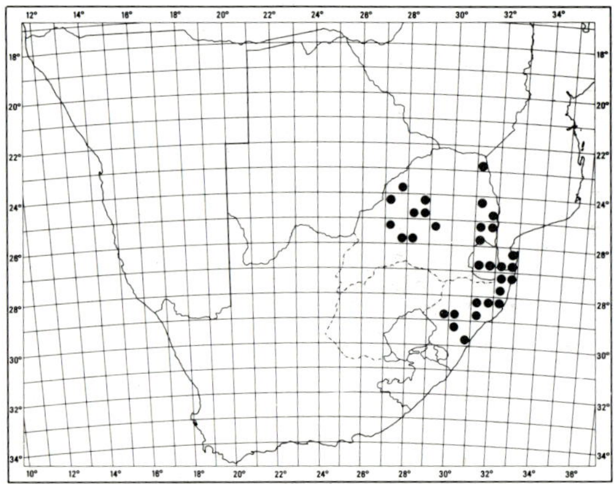

MAP 12.— **Commiphora neglecta**

C. neglecta differs from related species in its particular combination of characteristics rather than in any single outstanding feature. This is probably the reason why it was described only in 1951.

Although plants in nature only develop trifoliolate leaves, it has been observed that leaves of pole cuttings can be pinnate with two pairs of lateral leaflets.

Common name: Sweet-root Corkwood.

11. **Commiphora mollis** (*Oliv.*) *Engl.* in A. DC., Monogr. Phan. 4: 23 (1883); in Pflanzenfam. edn 2,19a: 435 (1931); Burtt Davy, Fl. Transv. 2: 484 (1932); Brenan in Kew Bull. 1950: 367 (1950); Exell & Mendonça in C.F.A. 1: 298 (1951); in F.Z. 2: 273 (1963); Merxm. in F.S.W.A. 23: 7 (1968); J. J. A. v.d. Walt in Bothalia 11: 73, fig. 39—44 (1973); in Mitt. bot. StSamml., Münch. 12: 211, fig. 8 (1975). Type: Mozambique, Chiramba, between Tete and coast, *Kirk* s.n (K, holo.!).

Balsamodendrum molle Oliv. in F.T.A. 1: 326 (1868). *Balsamea mollis* (Oliv.) Engl. in Bot. Jb. 1: 42 (1881).

Commiphora welwitschii Engl. in A. DC., Monogr. Phan. 4: 22 (1883); in Pflanzenfam. edn 2,19a: 435 (1931); Type: Angola, Huila, *Welwitsch* 4493 (G, holo., only photo seen; LISU!).

C. cinerea Engl. in Bot. Jb. 19: 139 (1895); in Pflanzen-fam. edn 2,19a: 435 (1931). Type: S.W.A./Namibia, Otji-tambi, *Gürich* 21 (B, holo.†; K, fragment!).

C. stuhlmannii Engl., Pflanzenw. Ost.-Afr. C: 230 (1895); in Pflanzenfam. edn 2,19a: 435 (1931); Type: Tan-zania, Bukombe, *Stuhlmann* 3450 (B, holo.†; K. fragment!; BM, sketch!).

C. dekindtiana Engl. in Bot. Jb. 34: 312 (1905); in Pflan-zenfam. edn 2,19a: 435 (1931). Type: Angola, Huila, Ben-guela, *Dekindt* 225 (B, holo.†; K, fragment!).

C. heterophylla Engl. in Bot. Jb. 34: 312 (1905); in Pflanzenfam. edn 2,19a: 435 (1931). Type: Tanzania, Kili-mandjaro area between Taveta and Bura, *Engler* 1906 (B, holo.†; K, fragment!).

C. montana Engl. in Bot. Jb. 34: 312 (1905); in Pflan-zenfam. edn 2,19a: 435 (1931). Type: Angola, Huila, Ben-guela, *Dekindt* 46 (B, holo.†; K, fragment!).

C. krausei Engl. in Bot. Jb. 44: 152 (1910); Pflanzen-fam. edn 2,19a: 435 (1931). Type: Tanzania, Tabora, *Von Trotha* 8a (B, holo.†; K, fragment!).

C. iringensis Engl. in Bot. Jb. 44: 150 (1910); in Pflan-zenfam. edn 2,19a: 435 (1931). Type: Tanzania, Uhehe, *Spiegel* sub Amani Herbarium 2507 (B, holo.†; EA!).

C. boehmii Engl. in Bot. Jb. 48: 472 (1913); in Pflanzen-fam. edn 2,19a: 435 (1931). Syntypes: Tanzania, Gonda, *Böhm* 281 (B †; K, fragment, lecto.!); Salanda, *Fischer* 292 (B†; K, fragment!).

C. ndemfi Engl. in Bot. Jb. 54: 293 (1917); in Pflanzen-fam. edn 2,19a: 435 (1931). Type: Tanzania, Urambo, *Stolz* 1678 (B, holo.†; K!, lecto.!; Z!; P!).

Dioecious tree 3—8 m tall with a single trunk; bark brown to greyish green, usually peeling in thick discs; young branchlets spar-

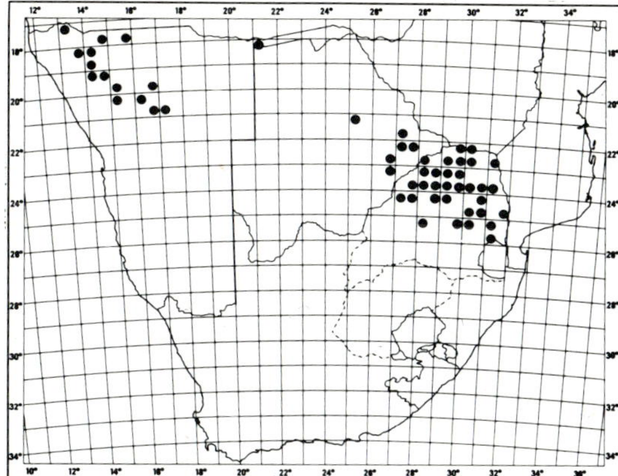

MAP 13.— **Commiphora mollis**

sely pilose to densely pubescent, not spine-tipped. *Leaves* pinnate, 2—6-jugate, occasionally trifoliolate, sparsely pilose to tomentose, greyish green but usually distinctly paler and more hairy below; petiole 10—30 mm long; leaflets elliptic or oblong-elliptic but terminal leaflet often obovate (7—)24(—52) × (4—)11 (—35) mm, sessile or subsessile but petiolule of terminal leaflet up to 15 mm long, margin entire, apex acute to obtuse, base cuneate to broadly cuneate. *Inflorescence*: simple or compound dichasial cymes up to 40 mm long, pilose to pubescent. *Flowers* unisexual, hypogynous. *Pedicel* 3—8 mm long, pedicel, calyx and corolla pilose to pubescent. *Disc* 4-lobed, not adnate to perianth. *Stamens* 8. *Fruit* subglobose, 15 × 13 × 13 mm, pilose to pubescent; putamen smooth; pseudo-aril with 4 winged arms, 2 commissural arms reaching almost to apex of putamen, 2 facial arms shorter.

Widely distributed in the northern part of S.W.A./Namibia, Botswana and northern Tvl. Grows on stony hills or well-drained sandy soil in savanna-woodland. Also recorded from Zimbabwe, Zambia, Malawi, Tanzania, Angola and Zaire. Map 13.

Vouchers: *De Winter* 7544; *Giess & Müller* 11801; *Meeuse* 10573; *Van der Walt* 77.

The variation in hairiness of the young stems and leaves, as well as the variation in the form and size of the leaflets, can account for the many synonyms.

This species is easily grown from pole cuttings which are often planted as fencing poles. The young branches are grazed by cattle and game.

Common name: Velvet Corkwood.

12. **Commiphora marlothii** *Engl.* in Bot. Jb. 44: 155 (1910); in Pflanzenfam. edn 2,19a: 438 (1931); Palgrave, Trees Cent. Afr. 55, t. &

photo (1956); Wild in F.Z. 2: 281 (1963); Lisowski, Malaisse & Symoens in Bull. Jard. bot. nat. Belg. 40: 360 (1970); J. J. A. v.d. Walt in Bothalia 11: 78 (1973). Syntypes: Zimbabwe, Matopos, *Marloth* 3397 (B†; PRE, lecto.!;K, fragment!); *Marloth* 3402 (B†).

Dioecious tree up to 9 m tall; bark peeling in large yellowish papery pieces; young branchlets obtuse, densely pilose to pubescent. *Leaves* pinnate, 3—5-jugate, pubescent to tomentose, dark green; petiole 15—95 mm long, with medullary vascular bundles; leaflets obovate to broadly elliptic; petiolules 1—2 mm long; margins crenate-serrate to finely lobed, apex obtuse to acute, base cuneate or rounded, terminal leaflet up to 80×40 mm, lateral leaflets up to 60×35 mm. *Inflorescence*: axillary, paniculate simple or compound dichasial cymes. *Flowers* unisexual, hypogynous. *Pedicel* less than 1 mm long, pedicel and calyx pubescent. *Petals* pilose outside. *Disc* 4-lobed, pilose, not adnate to perianth. *Stamens* 8. *Fruit* subglobose, ± 20×17×16 mm, pilose; putamen slightly rugose; pseudo-aril yellow, with 4 arms, 2 commissural arms reaching almost to apex of putamen, 2 facial arms shorter and of different lengths.

Widely distributed in northern Tvl. and also known from Botswana. It usually grows on arid mountain slopes or granite hills. Also recorded from Zimbabwe and Zambia. Map 14.

Vouchers: *Codd & De Winter* 5534; *De Winter* 7083; *Gerstner* 6043; *Van der Walt* 49;

The papery bark has been used as writing paper.

Common name: Paperbark Corkwood.

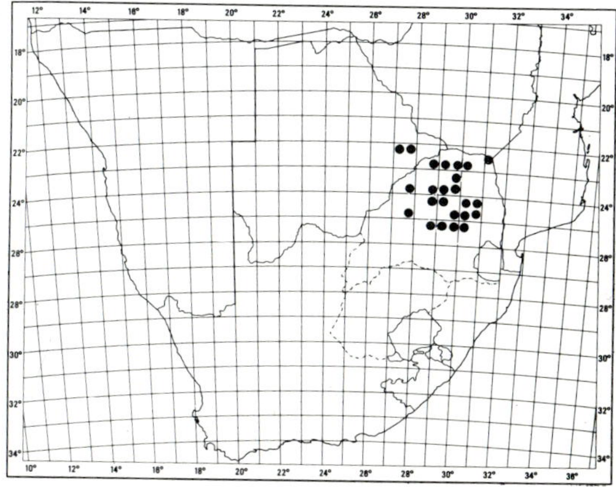

MAP 14.— **Commiphora marlothii**

13. Commiphora harveyi *(Engl.) Engl.* in A. DC., Monogr. Phan. 4: 25 (1883); in Pflanzenfam. edn 2,19a: 435 (1931); Burtt Davy, Fl. Transv. 2: 484 (1932); Henkel, Woody Pl. Natal 213 (1934); J. J. A. v.d. Walt in Bothalia 11: 76, fig. 45−50 (1973). Type: Natal, Durban, *Gerrard & McKen* 689 (TCD, holo.!; K!).

Balsamea harveyi Engl. in Bot. Jb. 1: 42 (1881).

Protium africanum Harv. in F.C. 2: 592 (1862); Swart, Monog. Protium 393 (1942).

Dioecious tree 4−18 m tall; bark peeling in large brown papery pieces or in thicker discs; young branchlets fluted and with a few very short hairs. *Leaves* pinnate, 2−3-jugate, occasionally trifoliolate, with a few very short hairs, green; petiole 10−60 mm long; leaflets lanceolate to elliptic to ovate; petiolules 5−15 mm long; margins crenate-serrate to coarsely crenate-serrate, apex acute, base cuneate, terminal leaflet up to 80×30 mm, lateral leaflets up to 60×25 mm. *Inflorescence*: axillary paniculate cymes. *Flowers* unisexual, hypogynous. *Pedicel* 2−3 mm long, pedicel and calyx with a few very short hairs. *Disc* 4-lobed, not adnate to perianth. *Stamens* 8. *Fruit* subglobose, ± 14×12×12 mm, glabrous, putamen smooth; pseudo-aril light red, with 4 arms, 2 commissural arms reaching almost to apex of putamen, 2 facial arms of variable length but shorter.

Occurs in north-eastern Tvl., eastern Tvl., Swaziland, Transkei and eastern Cape as far south as East London, but it is widely distributed in Natal and Zululand. It usually grows on the slopes of mountains or in kloofs as part of coastal forests. Also recorded from Mozambique. Map 15.

Vouchers: *Acocks* 12984; *Codd & De Winter* 3259; *Compton* 29733; *Van der Walt* 95.

C. harveyi is easily grown from pole cuttings which are often planted as fencing poles.

Common name: Red-stem Corkwood.

14. Commiphora mossambicensis *(Oliv.) Engl.* in A. DC., Monogr. Phan. 4: 26 (1883); Wild in F.Z. 2: 274, t. 51 fig. C (1963); in Mitt. bot. StSamml., Münch. 12: 214, fig. 10 (1975). Type: Malawi, Shire River, *Kirk* s.n. (K, holo.!).

Protium? mossambicense Oliv. in F.T.A. 1: 329 (1868).

Commiphora fischeri Engl. in Bot. Jb. 15: 97 (1893); Pflanzenfam. edn 2,19a: 435 (1931). Type: Tanzania, *sine loco, Fischer* 131 (B, holo.†; K, fragment!).

C. stolzii Engl. in Bot. Jb. 54: 292 (1917); Pflanzenfam. edn 2,19a: 435, fig. 203 (1931); Miller in Jl S. Afr. Bot. 18: 39 (1952). Type: Tanzania, Kyimbila, *Stolz* 1725 (B, holo.†; Z!; K!; BM!).

Dioecious tree 3−10 m tall; bark grey, smooth, not peeling; young branchlets sparsely pilose to densely pubescent with hairs and golden glandular hairs, not spine-tipped. *Leaves* trifoliolate or more rarely pinnate (2-jugate), sparsely pilose to densely pubescent with hairs and golden glandular hairs, bright green; petiole 50−80 mm long; leaflets ovate, broadly ovate, suborbicular or oblate, 15−70× 10−85 mm; petiolules 2−10 mm long; margin entire, apex often abruptly acuminate, base truncate. *Inflorescence*: thyrsoid, up to 70 mm long, sparsely pilose to densely pubescent with hairs and golden glandular hairs. *Flowers* unisexual, perigynous. *Pedicel* 1−2 mm, pedicel, calyx and corolla sparsely pilose to densely pubescent with hairs and golden glandular hairs. *Disc* with 4 large and 4 small lobes, upper part of lobes free and lower part adnate to

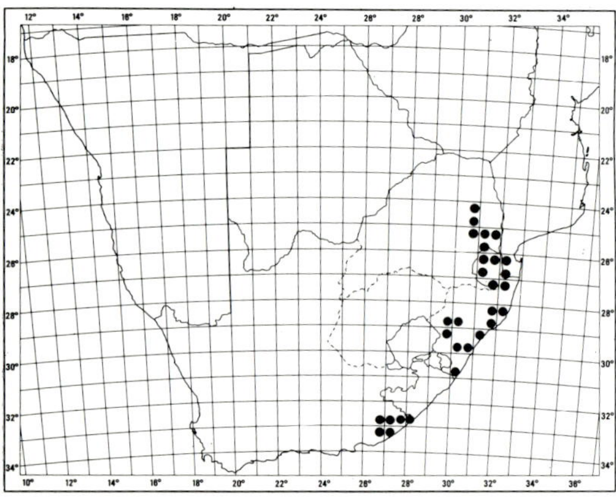

MAP 15.— **Commiphora harveyi**

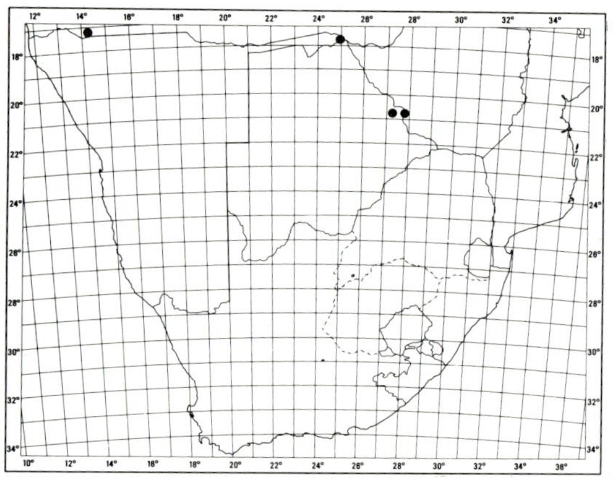

MAP 16.— **Commiphora mossambicensis**

hypanthium. *Stamens* 8. *Fruit* subglobose, ± 12×11×8 mm, sparsely pilose; putamen smooth; pseudo-aril with 2 commissural arms reaching almost to apex of putamen and 2 facial arms reaching three-quarters of the way upwards.

Within the F.S.A. area *C. mossambicensis* is known from Mpilila Island in the Caprivi strip and Botswana. Also recorded from Kenya, Malawi, Mozambique, Tanzania, Zambia and Zimbabwe. Map 16.

Vouchers: *Killick & Leistner* 3365; *Leach & Noel* 6; *Pole Evans* 3233.

Common name: Pepper-leaf Corkwood.

15. **Commiphora dinteri** *Engl.* in Bot. Jb. 44: 151 (1910); in Pflanzenfam. edn 2,19a: 438 (1931); Merxm. in F.S.W.A. 70: 6 (1968); J. J. A. v.d. Walt in Madoqua ser. 1: 9, t. 7−10 (1974); in Mitt. bot. StSamml., Münch. 12: 215, fig. 11, 28c & c_1, 29d & d_1 (1975). Syntypes: S.W.A./Namibia, Omatako, (sphalm. Omalako) *Dinter* 1393 (B,†); *sine loco, Dinter* 1477 (B,†; Z!; K, fragment!).

Dioecious shrub 0,5−3 m tall, many relatively thin stems sprouting from ground level; bark yellowish green to greyish brown with dark spots, smooth, not peeling; young branchlets glabrous, usually slender, not spine-tipped. *Leaves* trifoliolate, glabrous, green; petiole 2−18 mm long; leaflets obovate or broadly elliptic, subsessile, margin usually finely crenate-serrate, apex obtuse, seldom truncate or retuse, base cuneate, terminal leaflet (6−)14 (−22)×(4−)10(−15) mm, lateral leaflets (4−) 9(−12)×(2−)6(−9) mm. *Inflorescence*: flowers solitary. *Flowers* unisexual, perigynous, subsessile. *Calyx* glandular, otherwise glabrous. *Disc* 4-lobed, adnate to hypanthium. *Stamens* 8. *Fruit* ovoid, ± 11×8×8 mm, apiculate, glabrous; putamen smooth; pseudo-aril red, cupular with 4 (3) arms, 2 commissural arms reaching almost to apex of putamen, facial arms much shorter, arm on more convex face of putamen often completely undeveloped.

Occurs in S.W.A./Namibia, mainly in the Namib Desert or on its fringes. It is known from Orupemba southwards to the Zaris mountains near Maltahöhe, and eastwards to Rehoboth. Map 17.

Vouchers: *De Winter & Leistner* 5737; *Merxmüller & Giess* 935; *Van der Walt* 267.

The leaves of specimens collected near Orupemba are relatively large. Exceptionally large leaves of other species of *Commiphora* are also known from this part of Kaokoland.

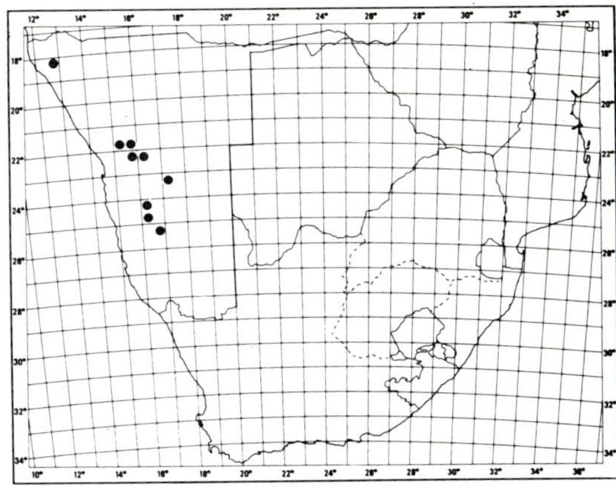

MAP 17.— **Commiphora dinteri**

Herbarium specimens of *C. dinteri* and *C. capensis* (no. 30) with leaves only, could be confused. The fruit of *C. dinteri*, however, has a pseudo-aril which is completely lacking in *C. capensis*.

Common name: Namib Corkwood.

16. **Commiphora gracilifrondosa** *Dinter ex J. J. A. v.d. Walt* in Jl S. Afr. Bot. 37: 190 (1971); in Bothalia 11: 95, fig. 94−99 (1973); in Mitt. bot. StSamml., Münch. 12: 216, fig. 12 (1975): Type: S.W.A./Namibia, Warmbad, near Auros, *Dinter* 5124 (BOL, holo.; S!; K!; B†).

Dioecious shrub-like tree 1−3 m tall, trunk branching above ground level into thick stems with succulent appearance; bark reddish brown with dark spots, not peeling; young branchlets glabrous, slender, not spine-tipped. *Leaves* trifoliolate with the terminal leaflet often irregularly lobed, glabrous, green; petiole 4−20 mm long; leaflets linear to cultrate but very variable in size and form, sessile or subsessile, margin irregularly and rather coarsely dentate-serrate, apex obtuse to acute, base cuneate, terminal leaflet (15−)25(−45)× (10−)15(−20) mm, lateral leaflets (7−)20 (−35)×1−2 mm. *Inflorescence*: dichasial cymes up to 50 mm long, sparsely glandular, or flowers solitary. *Flowers* unisexual, perigynous. *Pedicel* 1−4 mm long, pedicel, calyx and hypanthium sparsely glandular. *Disc* 4-lobed, adnate to hypanthium. *Stamens* 4 only. *Fruit* subglobose to ellipsoid, ± 10×8×7 mm, glabrous; putamen smooth; pseudo-aril cupular with 2 long commissural arms, covering lower quarter of more convex face of putamen and lower half of other face.

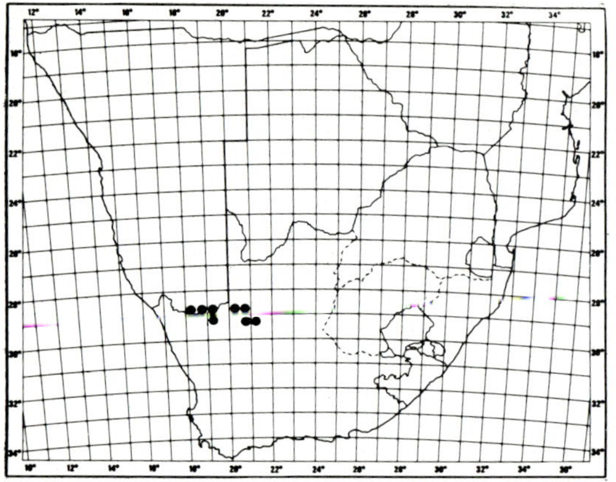

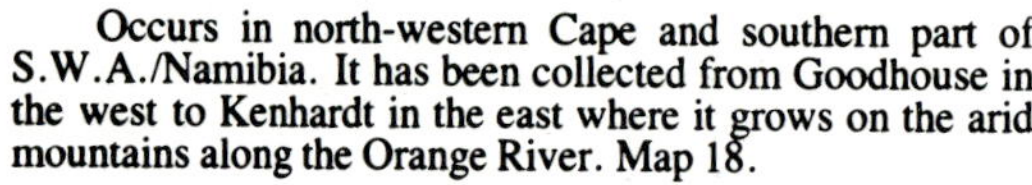

MAP 18.— **Commiphora gracilifrondosa**

Occurs in north-western Cape and southern part of S.W.A./Namibia. It has been collected from Goodhouse in the west to Kenhardt in the east where it grows on the arid mountains along the Orange River. Map 18.

Vouchers: *Acocks* 21795; *Giess & Müller* 12147; *Van der Walt* 124.

The relatively large leaves with characteristic leaflets, distinguish this species from others with a similar habit.

Young branches are browsed by goats and game. The local name of 'Suikerkan' is probably derived from the sweet taste of the wood.

Common name: Karree Corkwood.

17. **Commiphora oblanceolata** *Schinz* in Bull. Herb. Boissier, sér. 2, 8: 633 (1908); Engl. in Pflanzenfam. edn 2, 19a: 435 (1931); Merxm. in F.S.W.A. 70: 7 (1968); J. J. A. v.d. Walt in Jl S. Afr. Bot. 37: 196 (1971); in Madoqua ser. 1: 16, t. 27−29 (1974); in Mitt. bot. StSamml., Münch. 12: 217, fig. 13 (1975). Type: S.W.A./Namibia, Kaokoveld, Kan-Tal, *Dinter* 1497 (Z, holo.!; K!; W!).

Dioecious shrub-like tree, 1−3 m tall, trunk branching above ground level into thick stems with succulent appearance; bark grey to dark grey, smooth, not peeling; young branchlets glandular but otherwise glabrous, not spine-tipped. *Leaves* trifoliolate, glandular, green; petiole 3−25 mm long; leaflets narrowly oblanceolate to oblanceolate, (7−)13(−45)× (3−)4 (−9) mm, sessile or subsessile, margin finely serrate-dentate or almost entire, apex obtuse, base cuneate. *Inflorescence*: simple dichasial cymes up to 10 mm long, glandular, or flowers solitary. *Flowers* unisexual, perigynous. *Pedicel* 0,4−1 mm long, pedicel and calyx sparsely glandular. *Disc* 4-lobed, adnate to hypanthium.

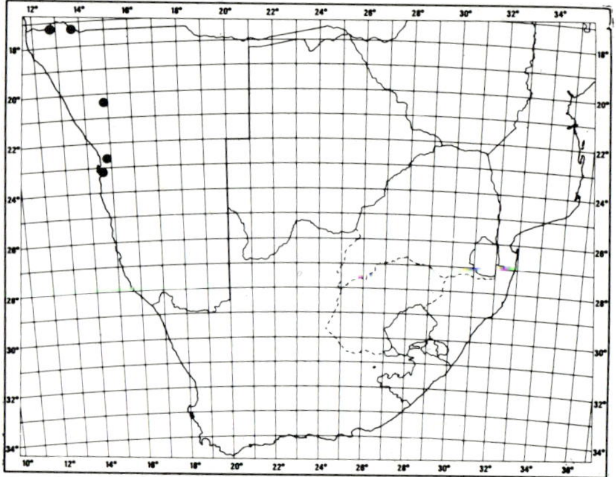

MAP 19.— **Commiphora oblanceolata**

Stamens 4 only. *Fruit* subglobose, ± 9×8×8 mm, glabrous; putamen and pseudo-aril not seen.

Occurs in the north-western part of S.W.A./Namibia from Swartbooisdrif in the north to the Namib Desert Park in the south. It is common on the farm Palmenhorst where it grows on the slopes of the arid mountains near the Swakop River. Map 19.

Vouchers: *Giess* 9342; *Kotze* 92; *Merxmüller & Giess* 1751.

C. oblanceolata and *C. gracilifrondosa* (no. 16) are the only two species with the reduced number of four stamens per flower. They can, however, easily be distinguished on leaf characteristics (Van der Walt, 1971).

The leaves of specimens collected in northern Kaokoland, are decidedly larger than those of the specimens occurring south.

Common name: Bastard Karree Corkwood.

18. **Commiphora namaensis** *Schinz* in Bull. Herb. Boissier, sér. 2, 8: 633 (1908); Merxm. in F.S.W.A. 70: 7 (1968); J. J. A. v.d. Walt in Bothalia 11: 92, fig. 88−93 (1973); in Mitt. bot. StSamml., Münch. 12: 218, fig. 14 (1975). Type: S.W.A./Namibia, Inachab, *Dinter* 958 (Z, holo.!; M, fragment!).

C. rotundifolia Dinter & Engl. in Bot. Jb. 46: 289 (1912); in Pflanzenfam. edn 2, 19a: 438 (1931). Type: S.W.A./Namibia, Seeheim, *Dinter* 1203 (B, holo.!; K, fragment!).

Dioecious shrub-like tree 0,5−3 m tall, trunk branching above ground level into many relatively thin stems; bark light grey, not peeling; young branchlets glabrous, not spine-tipped. *Leaves* simple but on long shoots occasionally trifoliolate with lateral leaflets a quarter to a third the size of terminal leaflet, glabrous, green; petiole 3−7 mm long; lamina rotund or

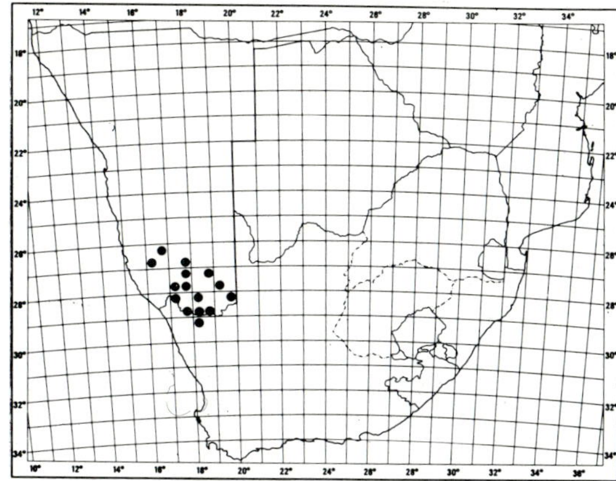

MAP 20.— **Commiphora namaensis**

orbicular, $(5-)9(-15)\times(4-)7(-12)$ mm, margin serrate-dentate, apex obtuse, base cuneate. *Inflorescence*: flowers in clusters. *Flowers* unisexual, perigynous, subsessile, glabrous. *Disc* 4-lobed, adnate to hypanthium. *Stamens* 8. *Fruit* subglobose or ellipsoid, $10\times8\times7$ mm, glabrous; putamen slightly rugose; pseudo-aril red, cupular with 2 commissural arms, covering the lower third of the more convex face of putamen and the lower half of the other face.

C. namaensis shows a disjunct distribution pattern. It has been collected in the Brandberg area, but its main centre of distribution is in the southern part of S.W.A./Namibia and north-western Cape where it grows in the mountains near the Orange River from Goodhouse westwards. This area is extremely dry and hot with an annual rainfall of less than 80 mm. Map 20.

Vouchers: *Acocks* 18168; *Merxmüller & Giess* 1688; *Van der Walt* 113, 120.

Trifoliolate leaves were seen on long shoots of plants growing near Uis. Similar leaves and transitional forms of simple to trifoliolate leaves also developed on plants cultivated at Stellenbosch.

Common name: Nama Corkwood.

19. **Commiphora glaucescens** *Engl.* in Bot. Jb. 10: 283 (1888); in Pflanzenfam. edn 2,19a: 435 (1931): Exell & Mendonça in C.F.A. 1,2: 302 (1951); Merxm. in F.S.W.A. 70: 6 (1968); J. J. A. v.d. Walt in Madoqua ser. 1: 12, t. 17−20 (1974); in Mitt. bot. StSamml., Münch. 12: 219, fig. 15, 28a & a₁, 29e & e₁ (1975). Type: S.W.A./Namibia, Usakos, *Marloth* 1306 (B†).

C. pruinosa Engl. in Bot. Jb. 26: 368 (1899); in Pflanzenfam. edn 2,19a: 435 (1931). Type: S.W.A./Namibia, Otyimbingue, *Ilse Fischer* 168 (B, holo.†; K, fragment!; BM, sketch!).

C. hereroensis Schinz in Bull. Herb. Boissier, sér. 2, 8: 632 (1908). Syntypes: S.W.A./Namibia, Potmine, Ubib, *Fleck* 447 & 742 (Z!).

Dioecious tree 2−8 m tall with a single trunk or shrub-like tree 1−2 m tall with a trunk branching near ground level; bark yellowish brown to reddish brown, peeling in papery pieces or discoid flakes of the same colour; young branchlets glabrous or sparsely pilose to densely pilose, not spine-tipped. *Leaves* simple, glabrous or sparsely pilose to densely pilose, glaucous or pale green, subsessile, elliptic or broadly elliptic seldom obovate, $(15-)40(-100)\times(8-)25(-60)$ mm, margin entire, apex usually truncate, seldom retuse or acute, base truncate or cuneate. *Inflorescence*: simple or compound dichasial cymes or thyrsoid, up to 80 mm long, glabrous or pilose to densely pilose, often with leaf-like bracts up to 10×7 mm. *Flowers* unisexual, perigynous. *Pedicel* 2−10 mm long, pedicel, calyx and corolla glabrous or sparsely pilose to pilose. *Disc* 8-lobed, upper part of lobes free but lower part adnate to hypanthium. *Stamens* 8. *Fruit* ellipsoid, ± $11\times6\times4$ mm, glabrous or pilose; putamen smooth; pseudo-aril red, cupular with 4 or 3 short lobes, covering lower quarter to third of putamen, 2 commissural lobes slightly longer than 2 facial lobes, lobe on more convex face of putamen in some cases undeveloped.

Widely distributed in the western part of S.W.A./Namibia. It occurs from the northern border of Kaokoland to Maltahöhe in the south, and it has been collected as far east as Grootfontein. Also recorded from Angola. Map 21.

Vouchers: *De Winter & Giess* 7103; *Merxmüller & Giess* 909; *Van der Walt* 236, 254.

Differences in habit, colour of the bark, size and degree of hairiness of the leaves, occur among representatives

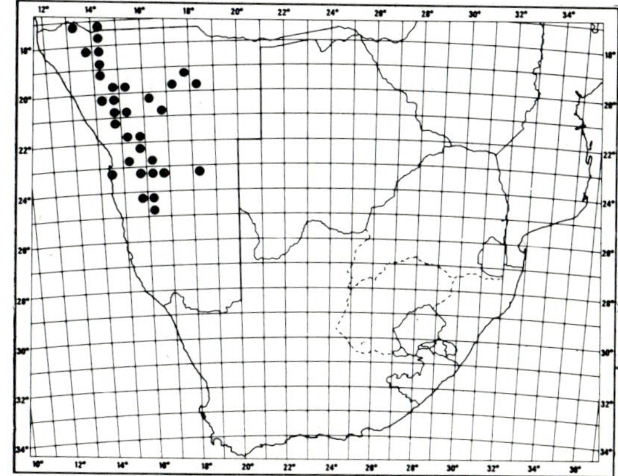

MAP 21.— **Commiphora glaucescens**

from different geographical areas. In the southern and eastern parts of its distribution range, *C. glaucescens* is a shrub with spreading, procumbent branches with small and glabrous leaves. In Kaokoland single-boled trees with large and very hairy leaves are found.

Elephants and other game browse on the young shoots and branches.

Common name: Blue-leaved Corkwood.

20. Commiphora wildii *Merxm.*, in Mitt. bot. StSamml., Münch. 3: 609 (1960); in F.S.W.A. 70: 9 (1968); J. J. A. v.d. Walt in Madoqua ser. 1: 21, t. 37−39 (1974); in Mitt. bot. StSamml., Münch. 12: 222, fig. 17, 27b & b₁, 29g & g₁ (1975). Type: S.W.A./Namibia, Kaokoland, W. of Sanitatas, *Merxmüller & Giess* 1453 (M, holo.!; PRE!).

Dioecious shrub-like tree 1−2,5 mm tall, trunk branching above ground level into relatively thick stems; bark grey-brown, shiny, smooth, occasionally peeling in papery strips; young branchlets pubescent to densely pubescent, often very short and stout. *Leaves* pinnately lobed or divided, (1−)2−4-jugate, pubescent to densely pubescent, glaucous; petiole 3−15 mm long; leaflets usually asymmetrically obovate or elliptic, (7−)20(−25)×(4−)10 (−12) mm, sessile, acroscopic margin incised to rachis but basiscopic margin decurrent along rachis, margin entire, apex obtuse or emarginate, seldom acute. *Inflorescence*: simple or compound dichasial cymes or thyrsoid, up to 40 mm long, pilose to densely pilose, or flowers solitary. *Flowers* unisexual, perigynous. *Pedicel* 2−6 mm long, pedicel, calyx and corolla sparsely pilose to densely pilose. *Disc* 8-lobed, adnate to hypanthium. *Stamens* 8. *Fruit* ovoid

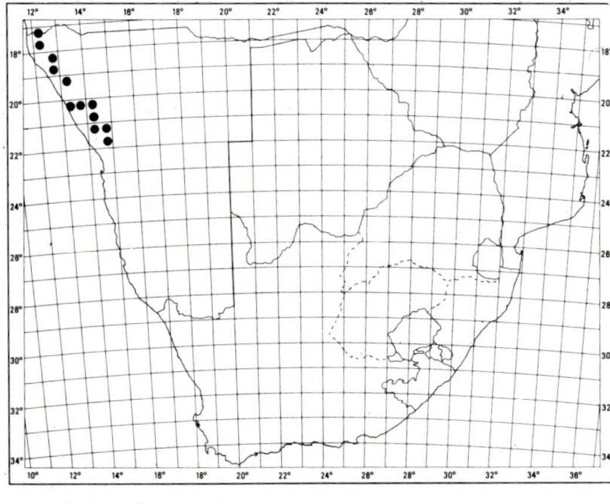

MAP 22.— **Commiphora wildii**

to subglobose, ± 11×10×9 mm, puberulous; putamen smooth; pseudo-aril yellow to orange, cupular with 2−4 short lobes, commissural lobes slightly longer than facial lobes, covering lower quarter of putamen.

Occurs in S.W.A./Namibia mainly on the edge of the Namib Desert, and it is found from the northern border of Kaokoland to Uis in the south. Also recorded from Angola. Map 22.

Vouchers: *De Winter & Hardy* 8170; *De Winter & Leistner* 5712; *Giess* 9714.

C. wildii is an attractive shrub with the glaucous leaves and shiny, grey-brown bark contrasting well against the black dolomite rocks on which it is often found. The leaves are characteristically lobed and resemble those of an oak.

Common name: Oak-leaved Corkwood.

21. Commiphora anacardiifolia *Dinter & Engl.* in Bot. Jb. 48: 475, fig. 2B−Bb (1912); Engl. in Pflanzenfam. edn 2,19a: 435 (1931); Merxm. in F.S.W.A. 70: 5 (1968); J. J. A. v.d. Walt in Madoqua ser. 1: 7, t. 1−3 (1974); in Mitt. bot. StSamml., Münch. 221, fig. 16, 29f & f1 (1975). Type: S.W.A./Namibia, Haobes, *Dinter* 1492 (B, holo.†; K, fragment and photo of holo.!; BM, sketch of holo.!).

Dioecious tree 5−10 m tall with a single trunk; bark yellowish brown, peeling in large papery pieces of the same colour; young branchlets pilose, obtuse. *Leaves* simple, pilose, dark green, sessile or subsessile, exceptionally large, (70−)130 (−200) × (30−) 80 (−140) mm, narrowly elliptic to broadly elliptic, midrib prominent, margin entire, apex obtuse, base cuneate. *Inflorescence*: thyrsoid, up to 200 mm long, many-flowered, pilose. *Flowers* unisexual, perigynous. *Pedicel* 2−4 mm long, pedicel, calyx and corolla sparsely pilose or pilose. *Disc* 8-lobed, upper part of lobes free but lower part adnate to hypanthium. *Stamens* 8. *Fruit* ovoid or obovoid, ± 15×11×10 mm, glabrous; putamen smooth; pseudo-aril yellow to orange, cupular with 4 short lobes, covering lower quarter to third of putamen.

S.W.A./Namibia, apparently confined to a small area in Kaokoland on the fringe of the Namib Desert, from the Sanitatas area in the north to Twyfelfontein in the south. Map 23.

Vouchers: *De Winter & Leistner* 5671; *Giess* 12610; *Merxmüller* 1434; *Van der Walt* 246.

The papery bark is typical of *Commiphora*, and the relatively large leaves are an outstanding characteristic to identify the species.

Common name: Large-leaved Corkwood.

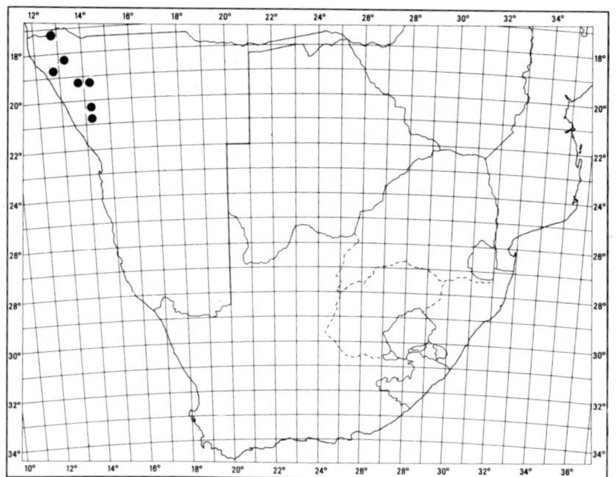

MAP 23.— **Commiphora anacardiifolia**

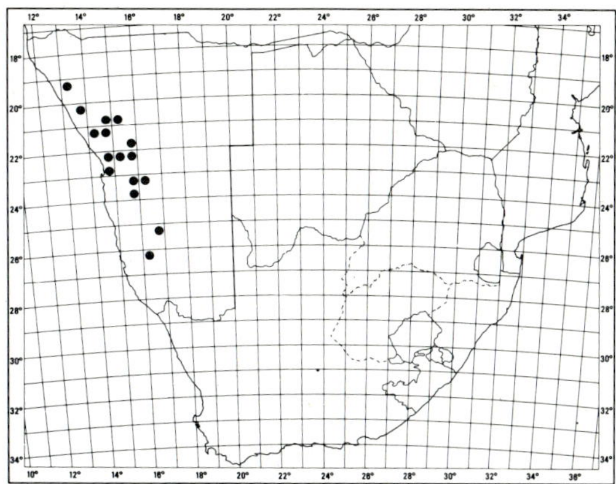

MAP 24.— **Commiphora saxicola**

22. Commiphora saxicola *Engl.* in Bot. Jb. 10: 283 (1888); in Pflanzenfam. edn 2,19a: 437 (1931); Merxm. in F.S.W.A. 70: 8 (1968); J. J. A. v.d. Walt in Madoqua ser. 1: 18, t. 30−33 (1974); in Mitt. bot. StSamml., Münch. 12: 224, fig. 19, 28b & b$_1$, 29h & h$_1$ (1975). Type: S.W.A./Namibia, Walvis Bay, *Marloth* 1221 (B, holo. †; PRE!).

C. dulcis Engl. in Bot. Jb. 19: 141 (1894). Type: S.W.A./Namibia, Tsoachoub near Husab, *Gürich* 6 (?).

Dioecious shrub-like tree or small tree 0,2−4 m tall with trunk 0,1−1,5 m long; bark grey, finely pitted, smooth, not peeling; young branchlets with glandular hairs but otherwise glabrous, not spine-tipped. *Leaves* pinnate, 2−6-jugate, with glandular hairs but otherwise glabrous, green; petiole 3−35 mm long; leaflets suborbicular to oblate, (3−)12(−17)×(3−)12(−17) mm; petiolules less than 1 mm long; margin rather coarsely crenate-serrate, apex emarginate or truncate, base cuneate or truncate. *Inflorescence*: raceme or thyrsoid up to 60 mm long, with glandular hairs. *Flowers* unisexual, perigynous, subsessile. *Pedicel,* calyx, corolla and ovary with glandular hairs. *Disc* obscurely 8-lobed, adnate to hypanthium. *Stamens* 8. *Fruit* oblong-ellipsoid, ± 10×6×6 mm, glabrous; putamen smooth; pseudo-aril orange, cupular without any lobes, covering lower third of putamen.

Occurs mainly on the fringe of the Namib Desert from Sesfontein in the north to Nauchas in the south. It grows on rocky hills or stony slopes but is also found on sandy flats. Map 24.

Vouchers: *De Winter* 3137; *De Winter & Hardy* 8167; *Merxmüller & Giess* 30668; *Van der Walt* 273.

The habit of *C. saxicola* is decidedly variable. Close to the west coast, north-east of Henties Bay, it is a small low-growing shrub with thick procumbent stems. More inland it is a single-boled tree attaining a height of up to four metres.

Common name: Rock Corkwood.

23. Commiphora edulis *(Klotzsch) Engl.* in A. DC., Monogr. Phan. 4: 22 (1883); in Pflanzenfam. edn 2,19a: 435 (1931); Wild in F.Z. 2: 279 (1963); J. J. A. v.d. Walt in Bothalia 11: 81, fig. 57−62 (1973); in Mitt. bot. StSamml., Münch. 12: 223, fig. 18 (1975). Type: Mozambique, Sena, *Peters* s.n. (B, holo. †; K!).

Hitzeria edulis Klotzsch in Peters, Reise Mossamb., Bot. 1: 89 (1861).

Commiphora chlorocarpa Engl. in Bot. Jb. 28: 414, t. IN (1901); in Pflanzenfam. edn 2,19a: 435 (1931). Type: Tanzania, Ruaha River, *Goetze* 452 (B, holo. †).

Dioecious many-stemmed shrub or small tree 2−6 m tall; bark light grey, smooth or flaking in small yellowish papery pieces; young branchlets densely pubescent, obtuse. *Leaves* pinnate, 2−6-jugate, pubescent, but scabrous above, greyish green; petiole 30−80 mm long, with medullary vascular bundles; leaflets narrowly elliptic to narrowly ovate, (30−)50(−65)× (20−)22(−30) mm; petiolules 4−10 mm long; margin usually entire, seldom finely crenate-serrate, apex acute or rounded, base obtuse. *Inflorescence*: compound dichasial cymes or thyrsoid up to 150 mm long, pubescent. *Flowers* unisexual, perigynous. *Pedicel* 1−1,5 mm long, pedicel, calyx and hypanthium pubescent. *Disc* much reduced without distinct lobes, adnate to hypanthium. *Stamens* 8. *Fruit* subglobose, ± 24×23×23 mm, pilose; puta-

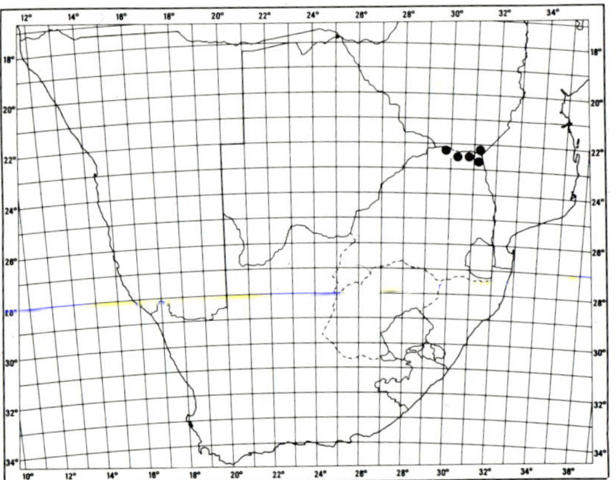

MAP 25.— **Commiphora edulis**

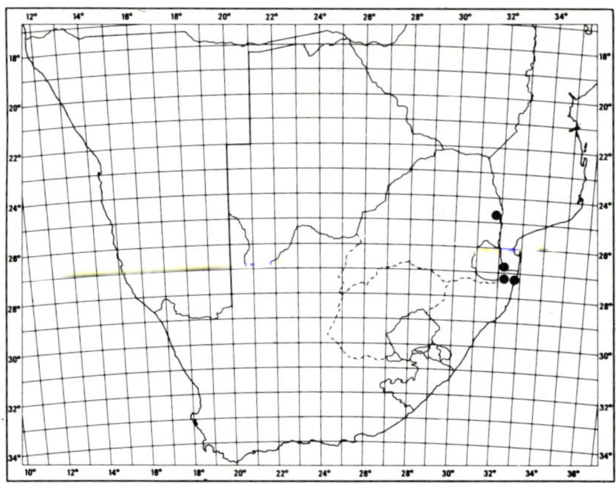

MAP 26.— **Commiphora zanzibarica**

men smooth; pseudo-aril red, cupular with 4 short lobes, covering lower third of putamen.

Occurs on Mpilila Island in the Caprivi strip (S.W.A./Namibia), Botswana and northern Tvl. where it is particularly common in the vicinity of the Limpopo River. It grows in savanna-woodland or broken mopaniveld in well-drained, sandy soil. Also recorded from Zimbabwe, Zambia, Malawi, Mozambique and Tanzania. Map 25.

Vouchers: *Codd & Dyer* 4653; *Killick & Leistner* 3337; *Van der Schijff* 5222; *Van der Walt* 24.

Gillett (pers. comm.) suggested that *C. edulis* is conspecific with *C. boiviniana* Engl. Several Kew specimens of the different subspecies of *C. boiviniana* were studied and compared with specimens of *C. edulis*. Differences in the size and shape of the leaflets exist between the subspecies of *C. boiviniana* and *C. edulis*, but the flowers and fruits have basically the same structure. Medullary vascular bundles which occur in the petiole of *C. edulis*, are also found in the petioles of all the subspecies of *C. boiviniana*. It seems likely therefore that *C. edulis* and *C. boiviniana* are conspecific.

C. edulis is one of the first *Commiphora* species of northern Transvaal to shed its leaves, the plants being leafless as early as March.

The fruits are eaten by birds, rodents and baboons.

Common name: Rough-leaved Corkwood.

24. **Commiphora zanzibarica** *(Baill.) Engl.* in A. DC., Monogr. Phan. 4: 28 (1883); in Pflanzenfam. edn 2,19a: 433 (1931); Wild in F.Z. 2: 279 (1963); J. J. A. v.d. Walt in Bothalia 11: 85, fig. 69−74 (1973). Type: Tanzania, Zanzibar, *Jablonski* s.n. (P, holo.!).

Balsamea zanzibarica Baill. in Adansonia 11: 180 (1874); Engl. in Bot. Jb. 1: 42 (1881).

Commiphora spondioides Engl. in Bot. Jb. 26: 371 (1899); in Pflanzenfam. edn 2, 19a: 433 (1931). Type: Mozambique, Maputo, *Schlechter* 11559 (B, holo.!; K!).

Dioecious tree, often many-stemmed and up to 7 m tall; bark grey, not peeling; young

branchlets shallowly fluted, glabrous. *Leaves* pinnate, 3−5-jugate, glabrous, green; petiole 15−60 mm long; leaflets oblanceolate to narrowly elliptic; petiolules 2−5 mm long; margins entire to finely serrate, apex obtuse to acute, base cuneate to rounded, terminal leaflet up to 70×35 mm, lateral leaflets up to 70×25 mm. *Inflorescence*: axillary, paniculate, simple or compound dichasial cymes. *Flowers* unisexual, perigynous. *Pedicel* 40−60 mm long, pedicel and calyx glabrous. *Disc* 4-lobed, adnate to hypanthium. *Stamens* 8. *Fruit* subglobose, ± 18×15×15 mm, glabrous; putamen smooth; pseudo-aril red, cupular, covering lower third to half of putamen.

So far this species has only been collected in Transvaal, near the Kumane Dam in the Kruger National Park and in Natal on the Makatini Flats in northern Zululand where it grows in deep sandy soil in savanna-woodland. Also recorded from Mozambique, Zimbabwe, Zanzibar, Tanzania and Kenya. Map 26.

Vouchers: *Garland* 314; *Van Wyk* 4874, 4917; *White* 10439.

Common name: Lebombo Corkwood.

25. **Commiphora woodii** *Engl.* in Bot. Jb. 15: 97 (1893); in Pflanzenfam. edn 2,19a: 435 (1913); J. J. A. v.d. Walt in Bothalia 11: 83, fig. 63−68 (1973). Syntypes: Natal, Durban, *Wood* sub NH 861 (B†?; BM, lecto.!); Pinetown, *Rehmann* s.n. (B†?); Inanda, *Rehmann* s.n. (B†?).

C. caryaefolia Oliv. in Hooker's Icon. Pl. 23: t. 2287 (1894); Engl. in Pflanzenfam. edn 2,19a: 435 (1931); Henkel, Woody Pl. Natal 213 (1934). Syntypes: Natal, Durban, *Wood* 4095 (NH, lecto.!; BOL!); Inanda, *Wood* 1046 (NH!); *Wood* 1409 (not seen); *Flanagan* 1107 (Z!).

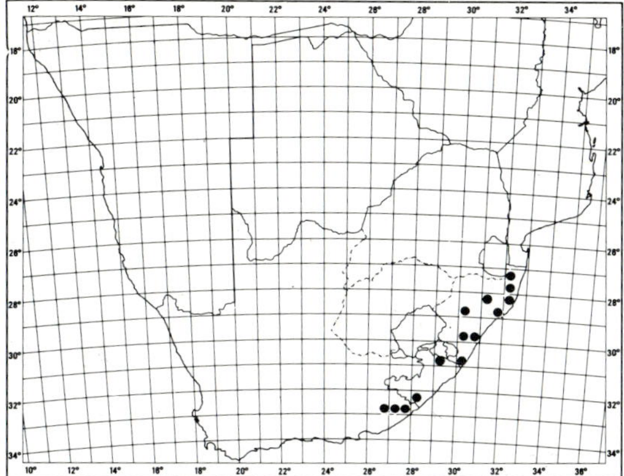

MAP 27.— **Commiphora woodii**

Dioecious tree up to 12 m tall; bark grey and not peeling; young branchlets shallowly fluted, obtuse, glabrous. *Leaves* pinnate, 3—5-jugate, glabrous, green; petiole 30—90 mm long; leaflets narrowly elliptic to elliptic; petiolules 2—5 mm long; margins crenate-serrate to coarsely crenate-serrate, apex acute, base cuneate to rounded, terminal leaflet up to 120×50 mm, lateral leaflets up to 130×50 mm. *Inflorescence*: axillary, paniculate, simple or dichasial cymes. *Flowers* unisexual, perigynous. *Pedicel* less than 1 mm long, pedicel and calyx glabrous. *Disc* 4-lobed, adnate to hypanthium. *Stamens* 8. *Fruit* subglobose, ± 20×19×18 mm, glabrous; putamen smooth; pseudo-aril red, cupular with 1 very short facial lobe, covering lower third of putamen. Figs 6, 7 & 8.

Occurs near the coast from Zululand southwards to East London. It usually grows on the slopes of mountains or in kloofs as part of coastal forests. Also recorded from Mozambique. Map 27.

Vouchers: *Codd* 1906; *De Winter & Vahrmeijer* 8481; *Moll & Morris* 663; *Pegler* 1137.

C. woodii and *C. zanzibarica* (no. 24), two closely related species, can be distinguished on features of the inflorescences, flowers and fruit. The inflorescences and flowers of *C. zanzibarica* are relatively long. Medullary vascular bundles occur in the petioles of *C. zanzibarica* but are absent in those of *C. woodii*. The leaves of *C. woodii* and *C. harveyi* (no. 13) are sometimes also confused. Very short hairs occur on the leaves of *C. harveyi* while those of *C. woodii* are glabrous.

C. woodii grows easily from pole cuttings which are often planted as fencing poles. Gum is prepared from the bark.

Common name: Forest Corkwood.

26. **Commiphora crenato-serrata** *Engl.* in Bot. Jb. 19: 140 (1894); in Pflanzenfam. edn 2,19a: 438 (1931); Merxm. in F.S.W.A. 70: 5 (1968); J. J. A. v.d. Walt in Madoqua ser. 1: 8, t. 4—6 (1974); in Mitt. bot. StSamml., Münch. 12: 226, fig. 20, 29i & i, (1975). Type: S.W.A./Namibia, Fransfontein, *Gürich* 71 (B, holo.†; BM, sketch of holo.!).

Dioecious tree 3—10 m tall with a single trunk; bark light grey to pale brown, pitted, smooth, not peeling; young branchlets with large glandular hairs but otherwise glabrous, conspicuously scarred, obtuse. *Leaves* pinnate, 3—7-jugate, with large glandular hairs on petiole, rachis and veins of leaflets but otherwise glabrous, green; petiole 40—80 mm long, with medullary vascular bundles; leaflets usually lanceolate, seldom narrowly lanceolate, (30—)60(—100)×(10—)20(—40) mm; petiolules 5—20 mm long; margin crenate-serrate, apex acuminate, base truncate. *Inflorescence* and flowers not seen. *Fruit* ovoid, ± 20×10×10 mm, conspicuously beaked, with a few glandular hairs; putamen smooth; pseudo-aril red, cupular with 0—4 short lobes, covering lower third to half of putamen.

Occurs mainly in Kaokoland from the Swartbooisdrif area in the north to Fransfontein in the south. It is also common on rocky outcrops of dolomite ridges near Otjovasandu in the Etosha National Park. Map 28.

Vouchers: *Hardy* 2082; *Tinley* 1249; *Van der Walt* 249.

C. crenato-serrata is often confused in the field with *Kirkia acuminata* Oliv. which it resembles superficially. The trees bear large quantites of fruit from December to April. A whitish-coloured resin with an aromatic odour is exuded when leaves or branchlets are picked.

Common name: Damara Corkwood.

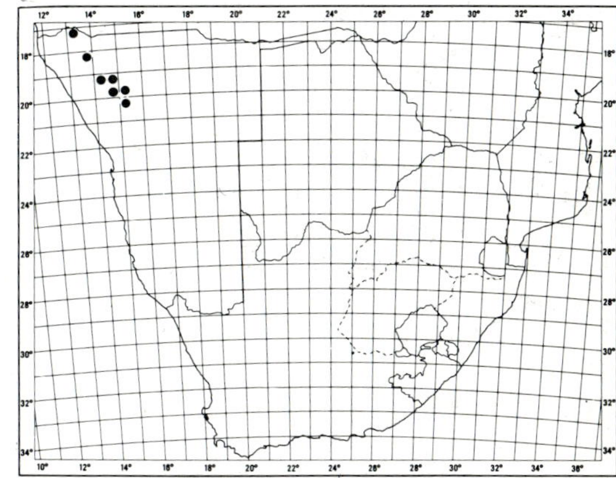

MAP 28.— **Commiphora crenato-serrata**

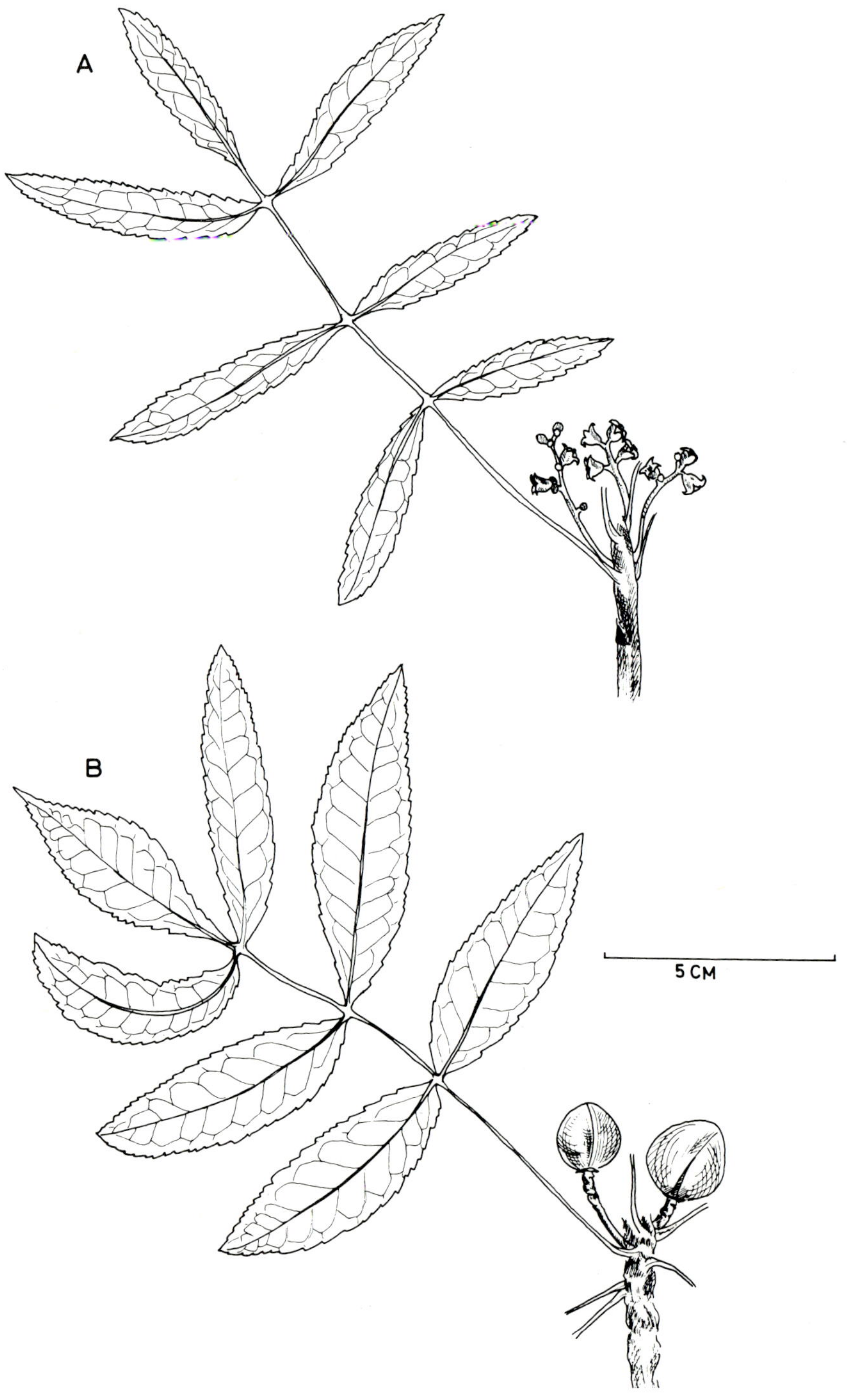

FIG. 6.—**Commiphora woodii:** A, branchlet with a leaf and flowers; B, branchlet with a leaf and mature fruits.

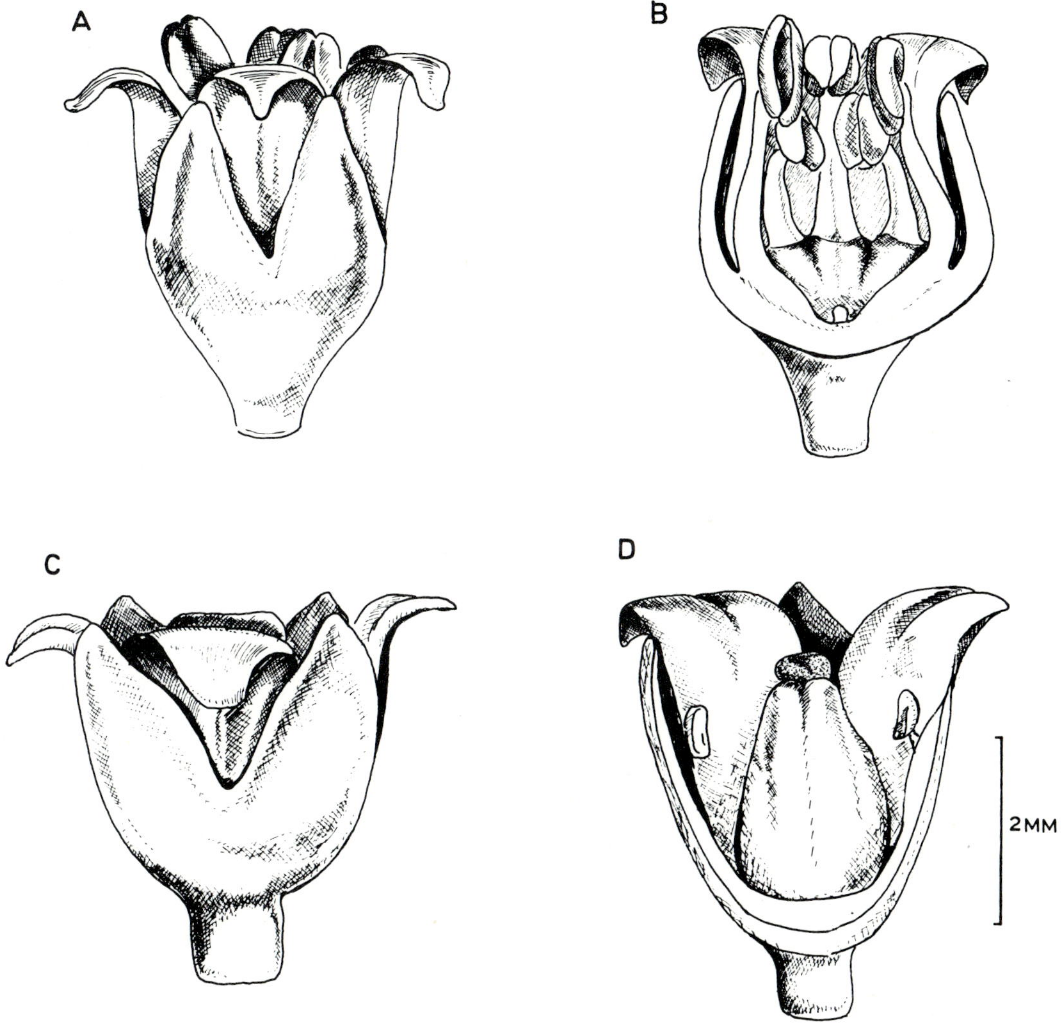

FIG. 7.—Flowers of **Commiphora woodii.** A, male flower; B, longitudinal section of male flower; C, female flower; D, female flower with the calyx and corolla partly removed.

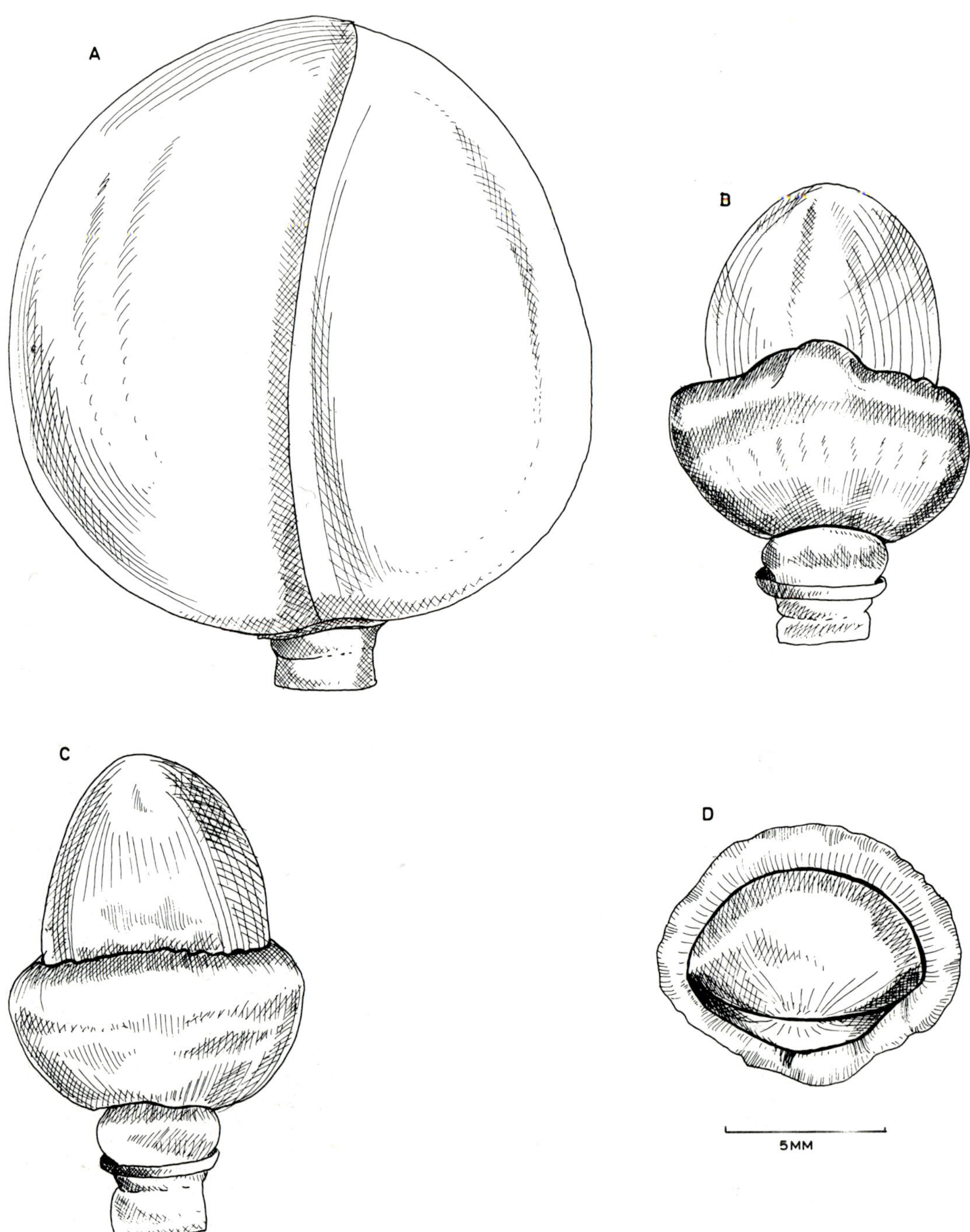

FIG. 8.—Fruit of **Commiphora woodii:** A, side view of the fruit; B, view of the less convex face of putamen with pseudo-aril; C, view of the more convex face of putamen with pseudo-aril; D, putamen and pseudo-aril as seen from above.

27. Commiphora tenuipetiolata *Engl.* in Bot. Jb. 48: 483, fig. 3L (1913); in Pflanzenfam. edn 2,19a: 438 (1931); Wild in F.Z. 2: 280 (1963); Merxm. in F.S.W.A. 70: 9 (1968); J. J. A. v.d. Walt in Bothalia 11: 87, fig. 75−81 (1973); in Mitt. bot. StSamml., Münch. 12: 227, fig. 21, 29j & j$_1$ (1975). Syntypes: S.W.A./Namibia, Otjiwarongo, Sesfontein, *Dinter* 1721 (B†; K, fragment!; BM, sketch!); S.W.A./Namibia, Bulspoort, *Dinter* 2109 (B†; K, fragment!).

Dioecious tree with a single trunk 2−12 m tall, occasionally shrub-like tree with a short trunk; bark peeling in yellowish white papery pieces to expose a blue-green underlayer or peeling in thick brownish discs; young branchlets glabrous, sparsely pilose but rarely pubescent, not spine-tipped. *Leaves* trifoliolate (then mostly glabrous) or pinnate, 2−4-jugate (then usually sparsely pilose to pubescent, occasionally also with glandular hairs), glaucous, yellowish green or green; petiole 7−50 mm long, usually relatively thin and slender especially in trifoliolate leaves; upper three-quarters of leaflet margin crenate-serrate, lower quarter subentire or entire, occasionally exclusively entire, terminal leaflet elliptic, broadly elliptic or obovate, 12−65×7−30 mm; petiolule 1−4(−12) mm long; apex acute or obtuse, base cuneate rarely obtuse, lateral leaflets elliptic to broadly elliptic rarely suborbicular, 6−45×4−25 mm, subsessile or sessile, apex acute rarely obtuse, base cuneate or obtuse. *Inflorescence*: simple or compound dichasial cymes up to 55 mm long, glabrous, sparsely pilose or pubescent. *Flowers* unisexual, perigynous. *Pedicel* 4−10 mm long, pedicel, hypanthium and calyx usually glabrous, seldom pilose, occasionally also with glandular hairs. *Disc* reduced, without distinct lobes, adnate to hypanthium. *Stamens* 8. *Fruit* subglobose, ± 15×13×10 mm, glabrous; pseudo-aril cupular with 2 facial lobes of variable length and form covering lower third to three-quarters of putamen; lobe on less convex face of putamen usually longer and larger than other lobe.

Occurs in the northern part of S.W.A./Namibia, Botswana and northern Tvl. where it is particularly common north of the Soutpansberg. Grows in sandy, well-drained soil. Also recorded from Zimbabwe. Map 29.

Vouchers: *Codd* 4836; *De Winter & Hardy* 8065; *Van der Walt* 61, 284.

It is usually almost impossible to distinguish between *C. angolensis* (no. 28) and hairy forms of *C. tenuipetiolata*, using leaf characters alone. The glabrous forms of *C. tenui-*

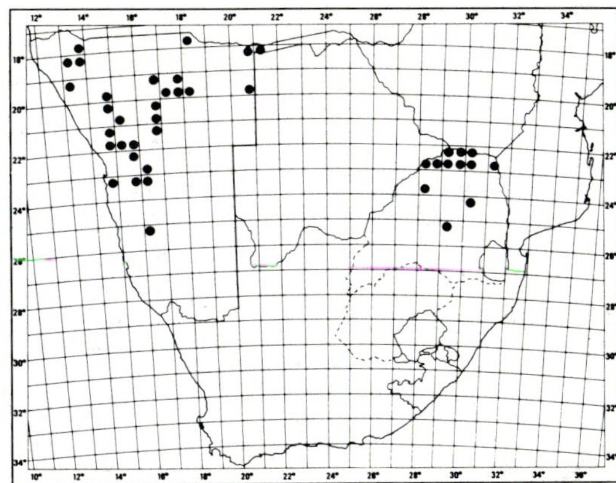

MAP 29.— **Commiphora tenuipetiolata**

petiolata usually have trifoliolate leaves with a thin and slender petiole, whereas the hairy forms have trifoliolate or pinnate leaves with a 'normal' petiole.

Common name: White-stem Corkwood.

28. Commiphora angolensis *Engl.* in A. DC., Monogr. Phan. 4: 24 (1883); in Pflanzenfam. edn 2,19a: 438 (1931); Exell & Mendonça in C.F.A. 1: 300 (1951); Wild in F.Z. 2: 281 (1963); Merxm. in F.S.W.A. 70: 5 (1968); J. J. A. v.d. Walt in Bothalia 11: 90, fig. 82−87 (1973); in Mitt. bot. StSamml., Münch. 12: 229, fig. 22, 29k & k$_1$ (1975). Syntypes: Angola, Luanda, *Welwitsch* 4485 (LISU, lecto.!); *sine loco* 4488 (G, only photo. seen; LISU!).

Balsamea angolensis (Engl.) Hiern, Cat. Afr. Pl. Welw. 1: 24 (1896).

Commiphora oliveri Engl. in A. DC., Monogr. Phan. 4: 24 (1883); in Pflanzenfam. edn 2,19a: 438 (1931). Type: Botswana, *Baines* s.n. (K, holo.!).

C. rehmannii Eng. in A. DC., Monogr. Phan. 4: 15 (1883). Type: Transvaal, *Rehmann* s.n. (B, holo.†; K, fragment and photo. of holo.!; BM, sketch of holo.!); Klippan, *Rehmann* 5324 (Z, lecto.!).

C. longebracteata Engl. in A. DC., Monogr. Phan. 4: 19 (1883); in Pflanzenfam. edn 2,19a: 438 (1931). Type: Angola, *Welwitsch* 4494 (G, holo., only photo. seen; LISU!).

C. kwebensis N.E. Br. in Kew Bull. 1909: 98 (1909). Syntypes: Botswana, Kwebe Hills, *Lugard* 34 (K!); *Lugard* 86 (K, lecto.!).

C. gossweileri Engl. in Bot. Jb. 44: 147 (1910). Type: Angola, Luanda, *Gossweiler* 442 (B, holo.†; K, lecto.!; BM!).

C. nigrescens Engl. in Bot. Jb. 44: 148 (1910); in Pflanzenfam. edn 2,19a: 438 (1931). Syntypes: S.W.A./Namibia, Grootfontein, *Dinter* 727 (B,†; K, fragment!); *Dinter* 727a (B†; BM, sketch, lecto.!).

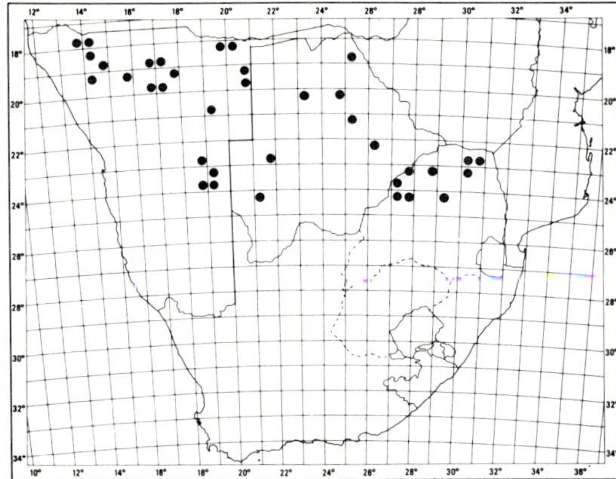

MAP 30.— **Commiphora angolensis**

Dioecious many-stemmed shrub or bush 0,5−2,5 m tall; bark grey to dark grey with brownish lenticels, often flaking locally in yellowish papery pieces; young branchlets sparsely pilose to densely pubescent, not spine-tipped. *Leaves* trifoliolate or pinnate, 2−4-jugate, sparsely pilose to densely pubescent, green; petiole 5−22 mm; upper three-quarters of leaflet margin crenate-serrate, lower quarter subentire to entire, terminal leaflet elliptic to broadly elliptic rarely obovate, (10−)16(−32)×(7−)10(−20) mm; petiolule 1−3(−9) mm; apex acute, rarely obtuse, base cuneate, rarely obtuse, lateral leaflets elliptic to broadly elliptic, rarely suborbicular, (5−)12(−30)×(4−)8(−18) mm, subsessile or sessile, apex acute, rarely obtuse, base cuneate or obtuse. *Inflorescence*: simple or compound dichasial cymes up to 50 mm long, sparsely pilose to densely pubescent. *Flowers* unisexual, perigynous. *Pedicel* 2−5 mm long, pedicel, hypanthium, calyx and corolla sparsely pilose to densely pubescent. *Disc* reduced without distinct lobes, adnate to hypanthium. *Stamens* 8. *Fruit* subglobose or ellipsoid, ± 13×11×9 mm, pilose; putamen smooth; pseudo-aril cupular with 2 facial lobes of variable length and form, covering lower quarter to half of putamen, lobe on less convex face of putamen usually longer than other lobe.

Widely distributed in the northern part of S.W.A./Namibia and also known from Botswana and a few localities in northern Tvl. It grows in deep sandy soil presumably derived from the Kalahari. Also recorded from Angola, Malawi, Zimbabwe and Zambia. Map 30.

Vouchers: *Codd* 5857; *De Winter* 3668; *Giess* 9742; *Van der Walt* 25.

In the Transvaal, *C. angolensis* and *C. tenuipetiolata* (no. 27), are two clearly delimited species, easily distinguishable by the indumentum of the branchlets and leaves.

The branchlets and leaves of *C. angolensis* are sparsely pilose to densely pubescent but those of *C. tenuipetiolata* are glabrous. In this area the two species may occur together, *C. angolensis* as a many-stemmed shrub and *C. tenuipetiolata* as a tree with a single trunk.

Near Ohopoho in S.W.A./Namibia these two species also occur together with the respective habits as in the Transvaal. The leaves of *C. tenuipetiolata* in this area, however, are hairy as those of *C. angolensis*.

Common name: Sand Corkwood.

29. **Commiphora kraeuseliana** *Heine* in Senckenberg. biol. 37: 493, fig. 1 & 2 (1956); Merxm. in F.S.W.A. 70: 7 (1968); J. J. A. v.d. Walt in Madoqua ser. 1: 14, t. 21−23 (1974); in Mitt. bot. StSamml., Münch. 12: 231, fig. 23, 29 (1975). Type: S.W.A./Namibia, Tafelberg near Petrified Forest, *Kräusel* 634 (FR, holo.!; M!).

Dioecious shrub 1−2 m tall with many relatively thin and slender stems sprouting from ground level; bark greyish brown or yellowish, peeling at the base of the stems in brownish papery pieces; young branchlets glabrous, scarred, relatively short and stout. *Leaves* pinnate, 6−8-jugate, glabrous, green; petiole 5−25 mm long; leaflets linear, (10−)15(−25) ×0,5−1 mm, sessile, margins entire. *Inflorescence*: thyrsoid, up to 60 mm long, villous, with large bracts up to 7×1 mm. *Flowers* unisexual, perigynous. *Pedicel* 4−8 mm long, pedicel, calyx and corolla villous or sparsely villous. *Disc* 8-lobed, adnate to hypanthium. *Stamens* 8. *Fruit* subglobose, ± 20×20×18 mm, glabrous; putamen smooth; pseudo-aril absent.

Occurs in the north-western part of S.W.A./Namibia. It has only been collected in the vicinity of the Brandberg and further north in Kaokoland. Map 31.

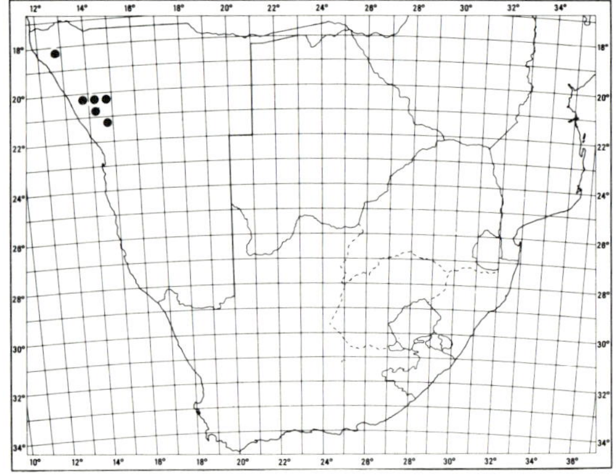

MAP 31.— **Commiphora kraeuseliana**

Vouchers: *De Winter & Leistner* 5719; *Giess* 3692; *Story* 5729.

The feathery leaves with linear pinnae are atypical of *Commiphora*. A very unpleasant odour is exuded when branchlets or fruits are picked. It has been reported that the seeds are eaten by local peoples.

Common name: Feather-leaved Corkwood.

30. **Commiphora capensis** *(Sond.) Engl.*
in A. DC., Monogr. Phan. 4: 18 (1883); Merxm. in F.S.W.A. 70: 5 (1968); J. J. A. v.d. Walt in Bothalia 11: 96, fig. 100−105 (1973); in Mitt. bot. StSamml., Münch. 12: 232, fig. 24 (1975). Type: N.W. Cape, between Natvoet and Orange River, *Drège* 6809 (ex parte) (S, holo.!; G!; MEL, fragment!).

Balsamodendrum capense Sond. in F.C. 1: 526 (1860).

Balsamea capensis (Sond.) Engl. in Bot. Jb. 1: 42 (1881).

Commiphora rangeana Engl. in Bot. Jb. 44: 149 (1910); in Pflanzenfam. edn 2, 19a: 438 (1913). Type: S.W.A./Namibia, Kovies Mountains, *Range* 172 (B, holo.†; BOL!).

C. ruquietiana Dinter & Engl. in Bot. Jb. 48: 482 (1913). Type: S.W.A./Namibia, Rotkop, *Dinter* 1023 (B, holo.†; SAM!).

Dioecious shrub-like tree 0,5−4 m tall, trunk branching near ground level into thick stems with succulent appearance; bark yellowish brown with dark patches, usually not peeling but occasionally flaking locally in small papery pieces; young branchlets glabrous, not spine-tipped. *Leaves* trifoliolate, glabrous, green; petiole 1−10 mm long; leaflets cordate, obovate or occasionally orbicular; petiolules 0,5−2 mm long; margin undulate or crenate, occasionally almost entire, apex emarginate or obtuse, base cuneate, seldom truncate, terminal leaflet (4−)9(−18)×(3−)8(−14) mm, lateral leaflets (3−)6(−13)×(2−)5(−10) mm. *Inflorescence*: simple dichasial cymes up to 10 mm long, glandular, or flowers solitary. *Flowers* unisexual, perigynous. *Pedicel* 0,5−1 mm long, calyx and hypanthium fleshy, calyx, hypanthium and corolla glandular. *Disc* 4-lobed, adnate to hypanthium. *Stamens* 8. *Fruit* ellipsoid, ± 12×10×6 mm, glabrous; putamen smooth; pseudo-aril absent.

Confined to the semi-desert areas of north-western Cape and southern S.W.A./Namibia. It grows in the mountains in the vicinity of the Orange River from Goodhouse westwards to the Richtersveld. These areas are extremely dry and hot with an annual rainfall of less than 80 mm. Map 32.

Vouchers: *De Winter & Hardy* 7918; *Giess* 2355; *Merxmüller & Giess* 3113.

C. capensis is closely related to *C. cervifolia* (no. 31) and they can be distinguished on leaf characteristics. As in

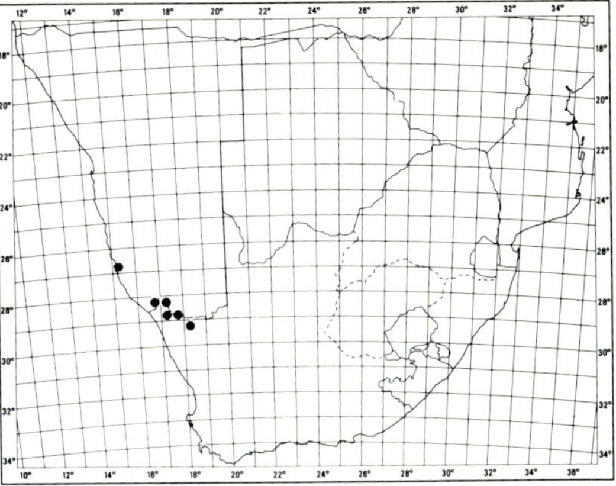

MAP 32.— **Commiphora capensis**

the case of *C. cervifolia*, but to a lesser extent, the living shoots on being touched, exude an aromatic secretion in such quantities that the stems become wet.

The fruits are eaten by animals.

Common name: Namaqua Corkwood.

31. **Commiphora cervifolia** *J. J. A. v.d. Walt* in Jl S. Afr. Bot. 37: 189 (1971); in Bothalia 11: 99, fig. 106−111 (1973); in Mitt. bot. StSamml., Münch. 12: 233, fig. 25 (1975). Type: N.W. Cape, 8 km S. of Viooolsdrif, *Van der Walt* 128 (PRE, holo.!; PRU!).

Dioecious shrub-like tree 0,5−3 m tall, trunk branching near ground level into thick stems with a succulent appearance; bark greyish green to yellowish brown with dark patches, not peeling; young branchlets glabrous, short and stout. *Leaves* trifoliolate, glabrous, green; petiole 2−5 mm long; leaflets cultrate or narrowly oblanceolate and usually irregularly lobed, (2−)7(−12)×(1−)2(−3) mm, sessile, margin entire irrespective of lobes, apex acute to obtuse, base cuneate. *Inflorescence*: dichasial cymes up to 10 mm long, glandular, or flowers solitary. *Flowers* unisexual, perigynous. *Pedicel* 1−1,5 mm long, calyx and hypanthium fleshy and glandular. *Disc* 4-lobed, adnate to hypanthium. *Stamens* 8. *Fruit* ellipsoid, ± 11×10×6 mm, glabrous; putamen smooth; pseudo-aril absent.

Confined to the semi-desert areas of southern S.W.A./Namibia and north-western Cape. It has been collected in the vicinity of Ai-Ais, at Goodhouse and near Viooolsdrif. The annual rainfall in these areas is less than 80 mm. Map 33.

Vouchers: *Van der Walt* 264, 304, 306.

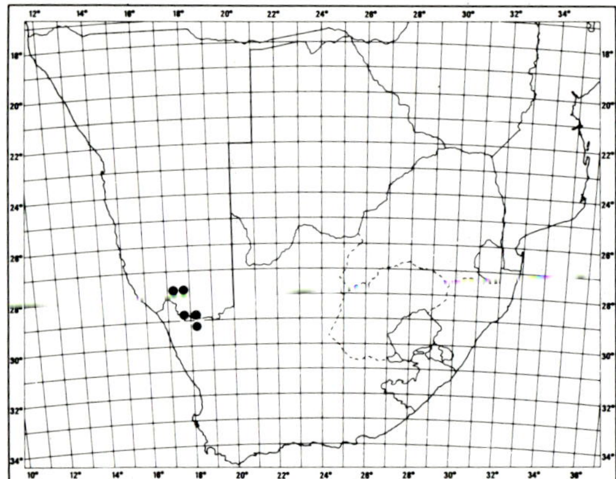

MAP 33.— **Commiphora cervifolia**

C. cervifolia is closely related to *C. capensis* (no. 30) and they have many characteristics in common, especially as far as habit and external features of stems and fruits are concerned. The leaflets of *C. capensis,* however, are rotund or obovate to cordate and not irregularly lobed such as those of *C. cervifolia.*

Living shoots, on being touched, exude an aromatic secretion in such quantities that the stems become wet.

Common name: Antler-leaved Corkwood.

Possible new species of *Commiphora* from S.W.A./Namibia

Merxmüller (1968) mentions four specimens (*De Winter & Leistner* 5670, *De Winter & Leistner* 5876; *Giess* 8921 and *Merxmüller & Giess* 1430) which resemble *C. saxicola* (no. 22) in some characters but differ from it in others. Duplicates of these specimens were examined, and the following deductions made:

(1) Medullary vascular bundles are lacking in the petioles of *C. saxicola* but they are present in the petioles of the specimens *Merxmüller & Giess* 1430 and *Giess* 8921. They are lacking in the petioles of *De Winter & Leistner* 5670 and 5876. It can therefore be concluded that the specimens *Merxmüller & Giess* 1430 and *Giess* 8921 represent a taxon other than *C. saxicola.*

(2) *Merxmüller & Giess* 1430 and *Giess* 8921 have medullary vascular bundles in the petioles, a character typical of *C. crenato-serrata* (no. 26). As Merxmüller (1968) points out, there is also a similarity between the leaves of *C. crenato-serrata* and those of *Merxmüller & Giess* 1430. The pseudo-aril of the latter specimen, however, has two short facial lobes and no commissural lobes, and is therefore not typical of *C. crenato-serrata.* More material is needed to solve this taxonomic problem.

(3) The specimens *De Winter & Leistner* 5670 and 5876 may belong to the same taxon. Both were collected from small trees with a grey, non-flaking bark, with leaflets of more or less equal size, and with glandular hairs. The obvious difference between the specimens lies in the shape of the leaflets.

Mendes (pers. comm.) expressed the opinion that the specimens *De Winter & Leistner* 5876 and *Menezes* 413 (from Angola, LISC) represent the same taxon and possibly a new species (*C. crassifoliolata* Mendes, nom. prov.).

Flowers and fruits of this supposedly new species are needed to determine whether it really differs from *C. saxicola.*

Merxmüller (1968) also mentions two specimens (*De Winter & Leistner* 5121 and *Giess, Volk & Bleissner* 6093) with typical and atypical characters of *C. crenato-serrata.* Mendes (pers. comm.) proposed that the specimens *De Winter & Leistner* 5121 and *Keet* 1622 (PRE) represent the same taxon and possibly a new species (*C. macrofoliolata* Mendes, nom. prov.). Also in this case, it is impossible to make a final decision on the validity of this proposed new species without flowering and fruiting material.

PTAEROXYLACEAE

by F. WHITE* and B. T. STYLES*

Dioecious, aromatic trees or shrubs with indistinct oil cavities in the younger parts. *Leaves* opposite, paripinnate, without stipules. *Flowers* unisexual, in contracted thyrses, actinomorphic. *Sepals* 4, free almost to the base, with open aestivation. *Petals* 4, free, imbricate. *Stamens* 4, free, alternating with the petals; filaments glabrous. *Disc* well-developed, broadly annular, intrastaminal. *Ovary* superior, laterally compressed, 2-locular, with axile placentation; ovules 1 per loculus, descending, with adaxial rhaphe; style about half as long as the ovary; stigmas 2, large, capitate, spreading. *Capsule* splitting into 2 persistent, bilobed valves; central column breaking up into a number of fibrous strands. *Seeds* with a long terminal wing.

A small family with two genera. The monotypic *Ptaeroxylon* is confined to the African mainland. *Cedrelopsis* Baill., with at least 7 species, occurs in Madagascar. It differs from *Ptaeroxylon* in having 5 sepals, 5 petals and 5 stamens and 3−5 bi-ovulate carpels.

Ptaeroxylon had been placed by most earlier workers in Sapindaceae or Meliaceae, more rarely in Rutaceae. The family Ptaeroxylaceae was tentatively suggested by Sonder (in F.C. 1: 242, 1860) and formally proposed by Leroy (in C.r. hebd. Séanc. Acad. Sci., Paris 248: 106, 1959, and in J. Agric. trop. Bot. appl. 7: 456, 1960).

The secondary xylem and pollen grains of *Ptaeroxylon* differ from all known Meliaceae. *Ptaeroxylon* also lacks limonoids, a group of tri-terpene derivatives which characterize Meliaceae and Rutaceae (D. A. H. Taylor, pers. comm.). The Ptaeroxylaceae share a few features of wood anatomy with Rutaceae and Sapindaceae and a few pollen characters with some Sapindaceae and Simaroubaceae. Their distinctive chromones of the ptaeroxylin group (Dean & Taylor in J. Chem. Soc. (C): 114, 1966), have been found in Cneoraceae and one genus each of Rutaceae and Simaroubaceae. It is likely that *Ptaeroxylon* and *Cedrelopsis* are distantly related to all these groups, but they differ in so many important respects that family rank is justified.

4157   PTAEROXYLON

Ptaeroxylon *Eckl. & Zeyh.*, Enum. 1: 54 (1834); Sond. in F.C. 1: 242 (1860); Harms in Pflanzenfam. edn 2, 19b1: 48 (1940); Exell & Mendonça in C.F.A. 1: 306 (1951); White & Styles in F.Z. 2: 547 (1966); Friedrich in F.S.W.A. 71: 3 (1968); R. A. Dyer, Gen. 1: 297 (1975).

Description as for the family.

One species confined to the southern half of Africa.

Ptaeroxylon obliquum *(Thunb.) Radlk.* in Sber. bayer. Akad. Wiss. 20: 165 (1890); Burtt Davy, Fl. Transv. 2: 487 (1932); Chalk *et al.* in Chalk & Burtt Davy, Forest Trees Timb. Br. Emp. 3: 56, fig. 8. 11 (1935); Exell & Mendonça in C.F.A. 1: 306 (1951); O. B. Miller in Jl S. Afr. Bot. 18: 39 (1952); White & Styles in F.Z. 2: 548, fig. 118 (1966); Ross, Fl. Natal 215 (1972); Palmer & Pitman, Trees S. Afr. 2: 1373 & photos (1972). Type: South Africa, *Thunberg* s.n. (S-THUNB.!).

Rhus obliqua Thunb., F.C. 2: 224 (1818) & edn Schult. 268 (1823).

Ptaeroxylon utile Eckl. & Zeyh., Enum. 1: 54 (1834); Harv., Thes. Cap. 1: 11, t. 17 ('1859', probably 1860); Sond. in F.C. 1: 243 (1860); Sim, For. Fl. Cape Col. 166, fig. 31 (1907). Type: South Africa, 'In forests by the Bushman's River and in the districts of Addo and Coega (Uitenhage)', *Ecklon* s.n. (K, in Herb. Benth., iso.!).

Ptaeroxylon utile forma *robustum* Szyszyl., Polypet. Disc. Rehm. 48 (1888). Type: South Africa, Transvaal, Houtbosch, *Rehmann* 6502 (Z, iso.!).

Kirkia ? lentiscoides Engl. in Bot. Jb. 32: 124 (1902). Type: Angola, Huila, *Antunes* 196 (B, holo., presumably destroyed; BM!; COI!).

Harrisonia lentiscoides (Engl.) Boas in Beih. bot. Zbl. 29: 341 (1913).

Shrub or tree ususally less than 15 m tall, occasionally up to 20 m or more, usually deciduous; bole up to 1 m or more in diameter. *Bark* whitish grey, smooth at first, later darker and with longitudinal fissures. *Leaves* up to 120 mm long; rhachis slightly winged, usually ending in a short appendage. *Leaflets* usually in 3−7 pairs, up to 50×24 mm, very asymmetric, apex obtuse, rounded or emarginate, rarely acute or mucronate; secondary nerves rather close together, prominent on both surfaces. *Inflorescence* up to 50 mm long, axillary or in axils of fallen leaves. *Flowers* pale yellow, usually appearing before or with the young leaves. *Calyx* 1 mm long, sparsely puberulous; lobes acute. *Petals* 5×1,5 mm, glabrous except for the ciliolate margin. *Stamens* 3,5 mm long;

---

* Department of Botany, University of Oxford, South Parks Road, Oxford OX1 3RA, United Kingdom.

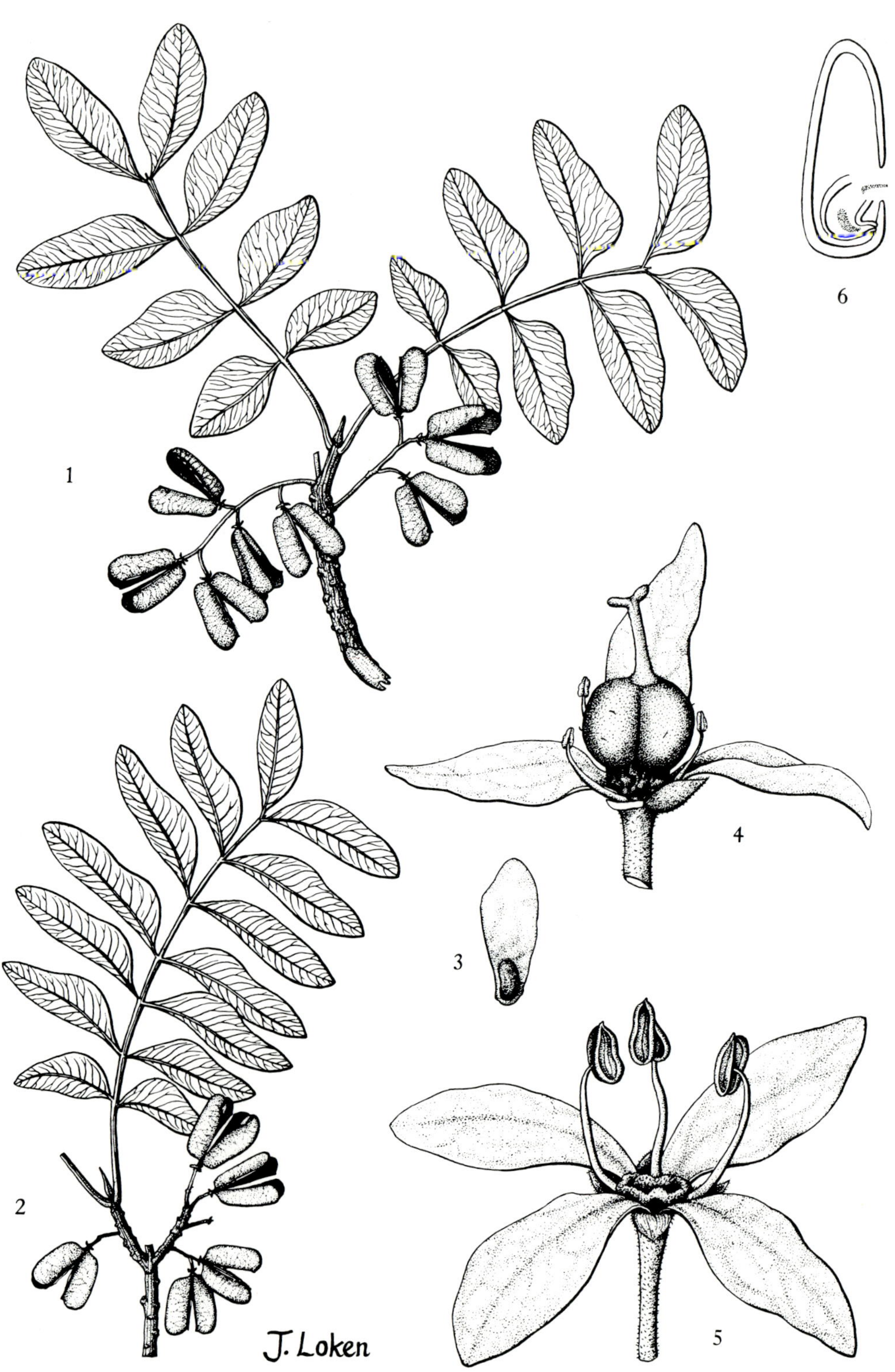
1
2
3
4
5
6
J. Loken

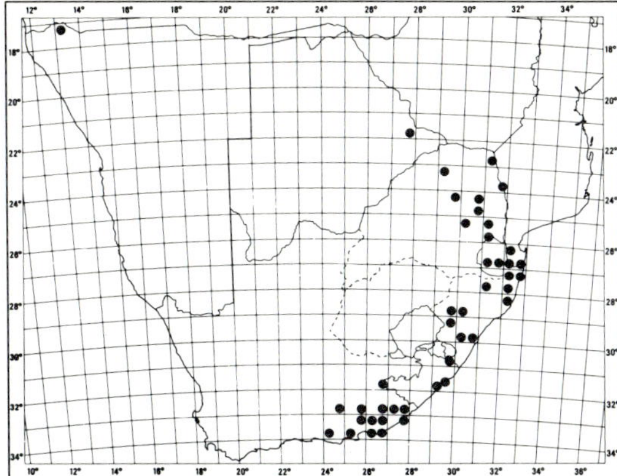

MAP 34.— **Ptaeroxylon obliquum**

staminodes 2 mm long; antherodes minute. *Ovary* 1,75 mm long, style about 1,25 mm long; pistillode minute, embedded in the disc, sometimes with 2 vestigial styles and loculi.

Capsule chestnut-brown with conspicuous veins, *c.* 18×12 mm. *Seed c.* 16×6 mm. Fig. 9.

Also in Mozambique, Zimbabwe, Angola and Tanzania. In southern Africa it occurs in S.W.A./Namibia, Botswana, Swaziland and in the Republic of South Africa from the northern Transvaal to the Cape. In various types of forest and bushland. From near sea-level to 1 370 m. Map 34.

Ptaeroxylon obliquum is appropriately known as 'Sneezewood' or 'Nieshout' as its fine sawdust provokes violent sneezing. The Zulu name is 'Umatati' and this name may have been bestowed on Tati District in the former Bechuanaland Prot. when invaded by the Matabele impis when, surprisingly, they found the tree so far from their native Zululand (O. B. Miller, loc. cit.). The wood, which was formerly of great economic importance, is strong and durable. It is harder and smoother than mahogany and of a finer grain, but it does not pick up as mahoganies tend to, and it turns very well. Its durability is its highest quality as it is practically imperishable. It has been greatly used for under-structure piles in house-building, fencing poles and telegraph poles. When used as machine bearings Sneezewood wears longer than brass or iron.

Vouchers: *Gibbs Russell* 4002; *Hall-Martin* 6666; *Story* 2786; *Taylor & Edwards* 8785; *Van der Walt* 343.

FIG. 9.—**Ptaeroxylon obliquum:** 1, branchlet with leaves and fruit, × 0,7 (*Armitage* 186/55); 2, branchlet with leaf and fruits, × 0,7 (*Barbosa* 721, from Maputo, Mozambique); 3, seed, × 1,3 (*S. Afr. For. Dept.* 5533); 4, female flower with one sepal and one petal removed, × 7 (*Galpin* 8097); 5, male flower with one stamen removed, × 7 (*Gomes e Sousa* 3869); 6, ovule in vertical section, × 10, after Mauritzon. By courtesy of the Editorial Board of *Flora Zambesiaca.*

MELIACEAE

by F. WHITE* and B. T. STYLES*

Trees, shrubs or shrublets; monoecious or dioecious, occasionally polygamous. *Indumentum* of simple, glandular and tufted-stellate hairs. *Leaves* usually spirally arranged, exstipulate, simply pinnate, 2–3-pinnate, 1–3-foliolate or simple. *Flowers* actinomorphic, bisexual or unisexual. *Sepals* small, 4–6, variously connate or almost free, never completely covering corolla in bud. *Petals* usually 4–5, free. *Stamens* usually 8 or 10 rarely 11–12, partly or completely fused to form a staminal tube, usually with appendages. *Ovary* superior, (2–)3–10(–20)-locular, with axile placentation and 1– many ovules per locule; style 1; style-head expanded, only partly stigmatic. *Fruit* a loculicidal or septifragal capsule or a drupe. *Seed* often with a fleshy aril or a dry membranous wing.

Characters not applicable in southern Africa: *Indumentum* rarely of peltate scales. *Leaves* very rarely decussate; leaf rhachis rarely ending in a terminal bud. *Inflorescence* sometimes cauliflorous, rarely epiphyllous. *Ovary* rarely unilocular. *Fruit* sometimes indehiscent and containing one or more arillate seeds, rarely a nut. *Seeds* sometimes buoyant and with a corky aril.

Genera 50, with about 550 species, almost confined to the tropics and subtropics. Six genera and 14 species are indigenous to southern Africa. *Melia azedarach* is extensively naturalized. Six other species which are planted for timber or ornament are briefly mentioned at the end of this account.

Further information on the taxonomy, chorology, ecology, reproductive biology and chemistry of South African Meliaceae can be found in a recent paper in Bothalia 16: 143 (1986).

1a Leaves simple:

  2a Stamens free almost to the base; fruit a large papery capsule, valves separating only near the apex.. 1. **Nymania**

  2b Stamens completely united or united almost to the apex; fruit a leathery or woody capsule, valves separating almost to the base ....2. **Turraea** (p.p.)

1b Leaves compound:

  3a Leaves bipinnate; staminal tube purplish; fruit a drupe.................................3. **Melia**

  3b Leaves once-pinnate:

    4a Leaves very variable and irregularly pinnate with deeply lobed or partite leaflets; fruit a capsule with arillate seeds ....2. **Turraea streyi**

    4b Leaves regularly pinnate with entire leaflets:

      5a Leaves imparipinnate; fruit a capsule with arillate seeds or a drupe:

        6a Staminal tube cup-shaped, filaments united to apex; fruit a drupe........ 4. **Ekebergia**

        6b Staminal tube elongate, filaments united only in upper half; fruit a capsule:

          7a Filaments with paired deltate appendages at apex; capsule with 3 leathery valves without appendages ....5. **Trichilia**

          7b Filaments without appendages; capsule with 5 woody valves covered with conspicuous ridges and antler-like appendages .... 6. **Pseudobersama**

      5b Leaves paripinnate; fruit a capsule with winged seeds .... 7. **Entandrophragma**

4168   **1. NYMANIA**

Nymania *Lindb.* in Not. Sällsk. Fauna Fl. Fenn. Förh. 9: 290 (1868); Harms in Pflanzenfam. edn 2, 19b1: 94, fig. 24 (1940); Friedrich in F.S.W.A. 71:3 (1968); R. A. Dyer, Gen. 1: 299 (1975); Pennington & Styles in Blumea 22: 460, fig. 4c (1975). Type species: *N. capensis* (Thunb.) Lindb.

Aitonia Thunb. in Physiogr. Sälsk. Handl. 1: 166 (1781) & in F.C. edn Schult. 508 (1823); Sond. in F.C. 1: 243 (1860), *nom. illegit.*, non *Aitonia* (‘*Aytonia*’) Forst. & Forst. f., Char. Gen. 147 (1776) (Marchantiaceae).

Aytonia L.f., Suppl. 49,303 (emend. *Aitonia* 468) (1781), *nom. illegit.*

---

* Department of Botany, University of Oxford, South Parks Road, Oxford OX1 3RA, United Kingdom.

FIG. 10.—**Nymania capensis:** 1, flowering twigs, × 0,7 (from *Comins* 1219); 1a, flowering twig, × 0,7 (from *Anderson* 3); 1b, flower in longitudinal section, × 3. (after Pennington & Styles, 1975); 1c, calyx lobe, × 3; 1d, lateral petal, × 1; 1e, cross section of fruit, × 1 (all after *Flower. Pl. Afr.*, fig. 1454); 1f, fruiting twigs, × 0,7 (from *Broom* s.n.); 1g, seed, × 2,7; 1h, seed in longitudinal section, × 2,7 (both after Harms, 1940).

Shrub. *Leaves* simple. *Indumentum* of simple hairs. *Flowers* bisexual, solitary, axillary. *Calyx* 4-lobed to near the base. *Petals* 4, free, imbricate. *Stamens* 8; filaments curved, fused only near the base, without appendages; anthers medifixed. *Disc* thin, partly fused to base of staminal tube. *Ovary* 4-lobed, with 4 locules each with 2 collateral ovules; style exserted with a simple, minute, scarcely expanded style-head. *Fruit* a papery capsule. *Seeds* reniform, minutely verrucose, puberulent.

A monotypic genus only known from southern Africa.

The name *Nymania* commemorates the Swedish botanist, C. F. Nyman, 1820–1893.

Nymania capensis *(Thunb.) Lindb.* in Not. Sällsk. Fauna Fl. Fenn. Förh. 9: 290 (1868); Verdoorn in Flower. Pl. Afr. 37, fig. 1454 (1965); Friedrich in F.S.W.A. 71: 3 (1968); Palmer & Pitman, Trees S. Afr. 2: 1059 & photos (1973). Type: South Africa, near Gouds River and Slang River, *Thunberg* s.n. (UPS, holo.).

Aitonia capensis Thunb. in Physiogr. Sälsk. Handl. 1: 166 with fig. (May 1781) & in F.C. edn Schult.: 508 (1823); Curtis's bot. Mag. fig. 173 (1792); Sond. in F.C. 1: 243 (1860).

Aitonia capensis var. *microphylla* Schinz in Verh. bot. Ver. Prov. Brandenb. 30: 156 (1888). Type: S.W.A./Namibia, Karakoes, *Schinz* s.n. (Z).

Rigidly branched, evergreen shrub 1–5 m high. *Leaves* up to 45×7 mm, 1-nerved, coriaceous, mostly in fascicles on abbreviated short shoots without internodes. *Flowers* dull red. *Calyx* lobes about 4×2,5 mm, puberulous. *Petals* erect, oblong-elliptic, about 15×10 mm, puberulous outside. *Stamens* up to 20 mm long, far-exserted. *Capsules* up to 40 mm diam., long-persistent on the plant, at first suffused with carmine, becoming straw-coloured and later silvery grey, deeply 4-lobed, with narrow, laterally compressed, wing-like locules. *Chromosome number*: 2n=40 *(Van Breda 4132, South Africa)*. Fig. 10.

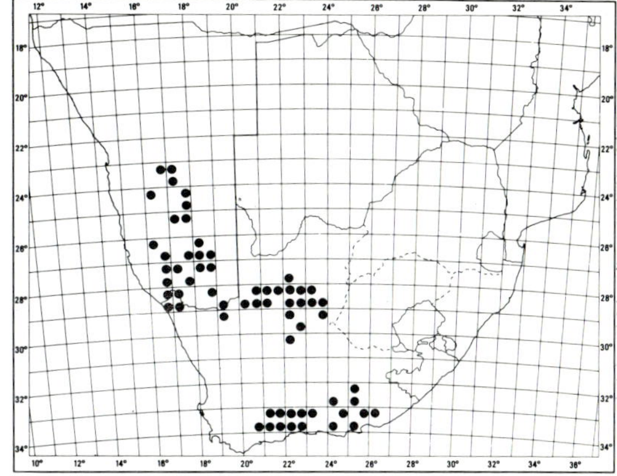

MAP 35.— **Nymania capensis**

N. capensis has a disjunct distribution. In the north it occurs from central S.W.A./Namibia southwards to the Orange River valley and from there inland almost as far as Kimberley. In the south it extends from the Little Karoo to the Great Fish River. In various types of karoid vegetation, often with succulents, and in bushland peripheral to the main area of the Karoo. Map 35.

Vouchers: *Acocks* 18176; *De Winter* 9111; *Leistner* 250; *Olivier* 3110; *Van Wyk* 754.

The Klapperbos or Chinese Lanterns is a striking and attractive plant because of its conspicuous red, inflated capsules. It is sometimes grown for ornament.

4171   **2. TURRAEA**

Turraea *L.*, Mant. 2: 150, 237 (1771); Sond. in F.C. 1: 244 (1860); Oliv. in F.T.A. 1: 330 (1868); C. DC. in Monogr. Phan. 1: 435 (1878); Harms in Pflanzenfam. edn 2, 19b1: 85, (1940); Staner & Gilbert in F.C.B. 7: 149 (1958); White & Styles in F.Z. 2: 307 (1963); R. A. Dyer, Gen. 1: 299 (1975); Pennington & Styles in Blumea 22: 455 (1975). Type species: *T. virens* L.

Trees, shrubs or suffrutices. *Leaves* simple, rarely compound. *Flowers* bisexual. *Calyx* cup-shaped with short teeth or with 5, almost free, foliaceous lobes. *Petals* 5, linear-spathulate or linear. *Staminal* tube cylindric, always with free or partly fused appendages; anthers 10, sessile or with short filaments. *Ovary* usually with 5 or 10 locules, each with 2, collateral or superposed ovules. *Style* expanded at the apex, usually exserted and then probably always functioning as a *receptaculum pollinis*, rarely included. *Fruit* a capsule with 5, 10 or 12 valves. *Seeds* reniform with a red or black, shining testa and a small conspicuous or inconspicuous aril.

Characters not applicable elsewhere: some species lack staminal appendages, or have ovaries with 20 locules, or a cup-shaped staminal tube or fewer than 5 petals.

J Loken
Rosemary Wise

About 50 species in Africa, Madagascar, the Mascarenes and Comores, and 1 widespread species in the tropical Far East.

Linnaeus named *Turraea* in honour of Antonio Turra, an Italian doctor and botanist, 1730–1796.

1a Calyx divided, free almost to the base, at least one third as long as the petals, lobes foliaceous; style shorter than the petals; subshrubs up to 0,4 m high:

  2a Leaves simple, deeply toothed or shallowly lobed in upper half; staminal tube split at apex.............. 5. *T. pulchella*

  2b Leaves compound; staminal tube undivided .... 6. *T. streyi*

1b Calyx cup-shaped with 5 short teeth, much shorter than the petals; style longer than the petals; shrubs or small trees:

  3a Petals up to 22 mm long; staminal tube suddenly expanded distally; bearded at the throat; appendages shorter than the anthers, connate to form an irregularly lobed frill beyond the insertion of the filaments:

    4a Branchlets of flowering specimens stout, usually more than 4 mm in diam.; bark on older branchlets becoming corky; flowers usually produced when the plant is leafless in subsessile fascicles on the older branchlets or at the ends of stout, slow-growing spur shoots densely beset with leaf scars. Leaf lamina mostly more than 100 mm long, lower surface densely pubescent, apex broadly acute, rounded or emarginate.......... 1. *T. nilotica*

    4b Branchlets of flowering specimens slender, usually less than 3 mm in diam.; bark not becoming corky; flowers usually produced when some leaves are present, occurring in leaf axils or in fascicles at the ends of slender shoots with widely spaced leaf scars. Leaf lamina up to 80 mm long, apex subacuminate, lower leaf surface glabrous or almost so... 2. *T. zambesica*

  3b Petals more than 30 mm long; staminal tube not suddenly expanded distally, not bearded at the throat; appendages longer than the anthers, free or connate only at the base:

    5a Leaf lamina up to 50 mm long, glabrous beneath, apex acute to emarginate; style straight; style-head turbinate or cylindric, widest at the (stigmatic) apex; capsule 5-valved, thinly woody .... 3. *T. obtusifolia*

    5b Leaf lamina 70–140 mm long, setulose beneath, especially on nerves and margin, apex acuminate; style bent at apex; style-head depressed-globose; capsule 10- or 12-valved, thickly woody..................... 4. *T. floribunda*

1. Turraea nilotica *Kotschy & Peyr.,* Pl. Tinn. 12, fig. 6 (1867); C. DC. in Monogr. Phan. 1: 445 (1878); White & Styles in F.Z. 2: 310, fig. 61d (1963). Syntypes: Sudan Republic, *De Heuglin*, 54, 55, 56 (W), *Knoblecher* s.n. (W).

Turraea randii Bak. f. in J. Bot., Lond. 37: 427 (1899); Burtt Davy, Fl. Transv. 2: 486 (1932). Type: Zimbabwe, Harare, *Rand* 562 (BM, holo.!).

Shrub or small tree up to 6 m tall, usually smaller. *Leaves* mostly elliptic or obovate. *Inflorescence* a 5–12-flowered, bracteate fascicle; bracts up to 5 mm long, subulate. *Petals* greenish white turning yellow with age. *Staminal tube* white, 10–15 mm long. *Capsule* 7 × 15 mm, 10-valved, depressed-globose, shallowly sulcate, glabrous, leathery or thinly woody. *Seeds* black, 5 × 3 mm, with a small orange or red aril completely concealed except for a horn-like extension which protrudes beyond the apex of the seed. *Chromosome number*: 2n = 50 (Zimbabwe). Fig. 11: 1.

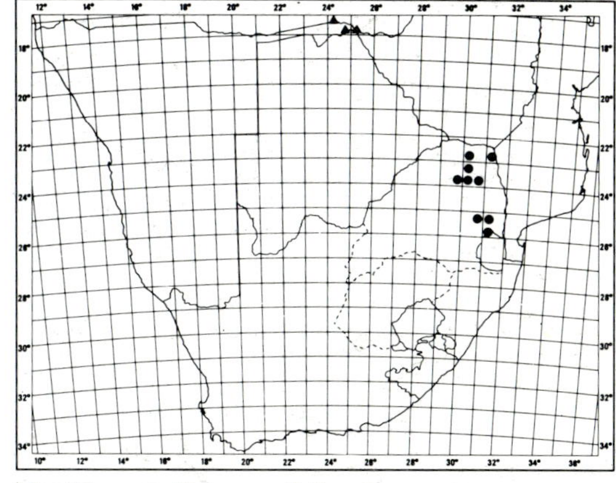

MAP 36.— ● **Turraea nilotica**
▲ **Turraea zambesica**

From the Sudan and Ethiopia southwards to the Transvaal. In the Transvaal it is rather rare and occurs in bushland and scrub woodland, mostly below 800 m, often on sandy soils. Map 36.

Voucher: *Codd* 1660.

FIG. 11.—1, **Turraea nilotica**: 1, leaves, × 0,5 (from *Robson* 665); 1a, flowering twig, × 0,5 (from *Bainbridge* 588); 1b, part of staminal tube and anthers, × 5,3 (from *Bainbridge* 588); 2, **T. zambesica:** 2, flowering twig, × 0,5 (from *Keay* FHI 21399); 2a, part of staminal tube and anthers (from *Miller* 377); 3, **T. obtusifolia:** 3, flowering twig, × 0,5 (from *Tyson* 30); 3a, calyx, × 5,3 (from *Meeuse* 9660); 3b, part of staminal tube and anthers, × 5,3 (from *Meeuse* 9660); 3c, style-head and stigma, × 5,3 (from *Meeuse* 9660); 3d, capsule, × 1,6 (from *Westfall* 1562); 4, **T. floribunda:** 4, flowering twig, × 0,5 (from *Wild* 4333); 4a, style-head and stigma, × 3,2 (from *Tyson* 10); 4b, fruit, × 0,8 (from Watt & Breyer-Brandwijk 1095); 4c, seed, × 2,1 (after *Flower. Pl. Afr.*, fig. 1499).

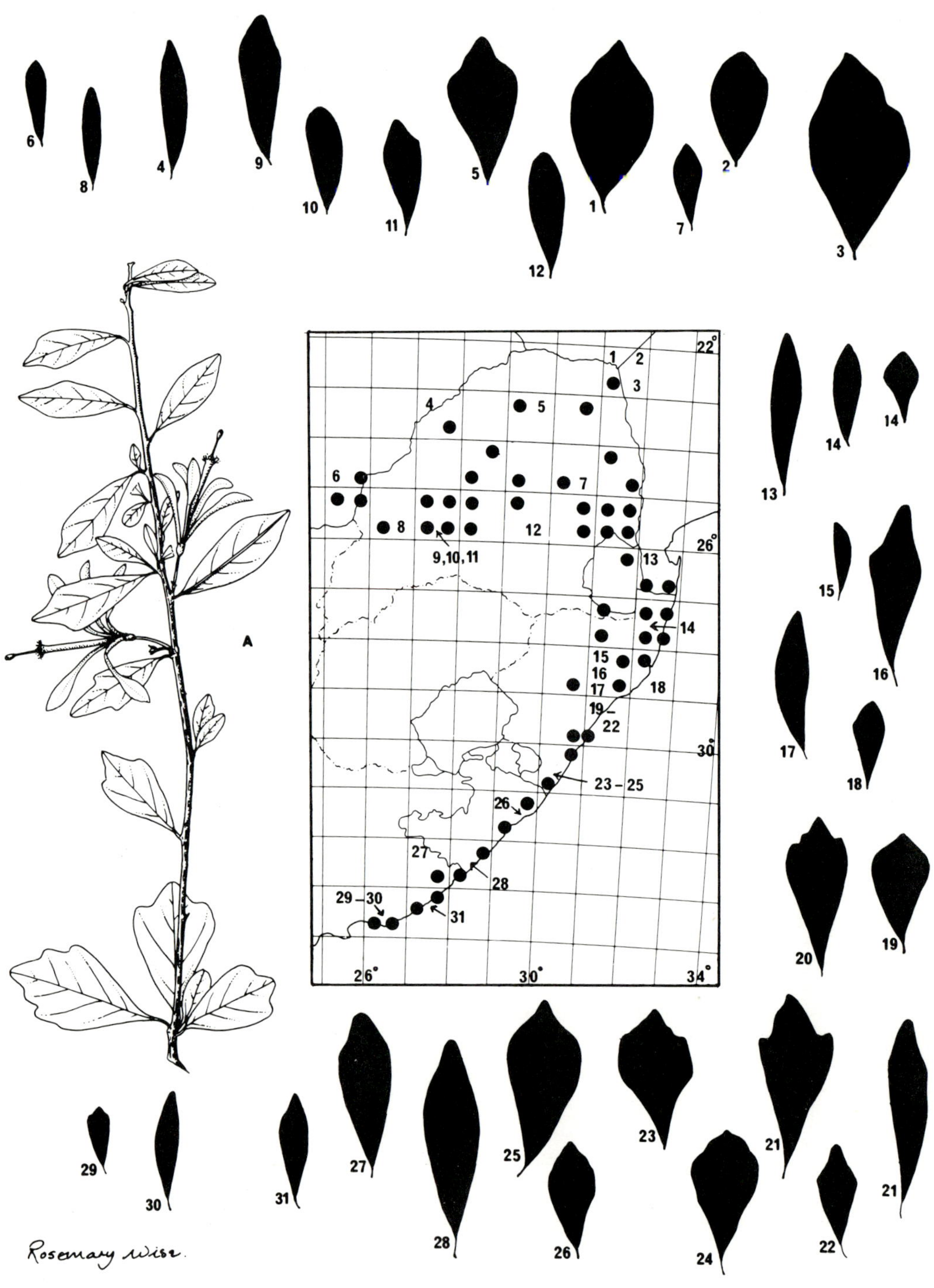

Rosemary Wise.

2. Turraea zambesica *Sprague & Hutch. ex Styles & White* in Bolm Soc. broteriana, sér. 2, 36: 71, fig. 1 (1962) & in F.Z. 2: 311 (1963); Sprague & Hutch. ex Hutch., Botanist in S. Afr. 481 (1946), English description only. Type: Zambia, Victoria Falls, *Hutchinson & Gillett* 3493 (K, holo.!; BM!; LISC!; SRGH!).

Turraea nilotica sensu Brem. & Oberm. in Ann. Transv. Mus. 16: 420 (1935); O. B. Miller in Jl S. Afr. Bot. 18: 40 (1952).

Closely related to *T. nilotica* (no. 1). In addition to the key characters, differing as follows: *Leaves* mostly rhombic-elliptic, rapidly tapering from near the middle to each end. *Staminal frill* usually densely ciliate. Fig. 11: 2.

Only known from the Zambezi Valley and its tributaries in the Caprivi Strip, Botswana, Zambia, Zimbabwe and Mozambique. In the Caprivi Strip and Botswana it is found in woodland and wooded grassland on the banks of the Zambezi and Chobe Rivers. Map 36.

Vouchers: *Codd* 7094; *Erens* 372; *Pole Evans* 4621; *P. A. Smith* 3935.

3. Turraea obtusifolia *Hochst.* in Flora 27: 296 [962] (1844); Sond. in F.C. 1: 245 (1860); Oliv. in F.T.A. 1: 331 (1868); C. DC. in Monogr. Phan. 1: 440 (1878); White & Styles in F.Z. 2: 313, fig. 61a (1963); Ross, Fl. Natal 216 (1972); Palmer & Pitman, Trees S. Afr. 2: 1063 & photo. (1973). Type: South Africa, Natal, *Krauss* 308 (G, iso.!; K, iso.!).

Turraea obtusifolia var. *microphylla* C. DC. in Monogr. Phan. 1: 440 (1878); Burtt Davy, Fl. Transv. 2: 486 (1932); O. B. Miller in Jl S. Afr. Bot. 18: 40 (1952). Type: South Africa, Port Alfred, near Barville Park, between Rietfontein and the seashore, *Burchell* 4106 (B, holo., presumably destroyed; K!; L!; M!; P!).

Turraea oblancifolia Brem. in Ann. Transv. Mus. 15: 245 (1933). Type: South Africa, between Warmbaths & Nylstroom, *Bremekamp & Schweickerdt* 4 (PRE, holo.!; K!).

Shrub 0,6−5 m high, sometimes scrambling. *Leaves* mostly in fascicles, lamina very variable, from narrowly oblanceolate and unlobed to cuneate-obovate and deeply lobed. *Inflorescence*: a 1−4-flowered axillary cyme. *Petals* pure white, narrowly spathulate, 25×3−45×5 mm. *Ovary* 5-locular, glabrous.

Capsule 6 × 12 mm, depressed-globose, shallowly sulcate. *Seeds* 5,5 × 3,5 mm, bright red; aril small, white, fleshy, almost confined to the adaxial depression and not visible from the outside. *Chromosome number*: 2n = 50 (Zimbabwe). Fig. 11: 3 & 12.

Only known from southern Africa from Botswana and Zimbabwe to southern Mozambique and the eastern Cape. In woodland, bushland and wooded grassland, especially among rocks; also in coastal thicket and inside and at the edges of various kinds of forest. In the Transvaal and Botswana, *T. obtusifolia* occurs mostly between 365 m and 1 200 m; further south between near sea-level and 400 m. Map (see fig. 12).

Vouchers: *Acocks* 8831; *Hansen* 3296; *Obermeyer* TM 35149; *Van der Schijff* 4000; *Venter* 4233.

4. Turraea floribunda *Hochst.* in Flora 27: 297 (1844); C. DC. in Monogr. Phan. 1: 445 (1878); Stapf in Curtis's bot. Mag. fig. 8944 (1923); White & Styles in F.Z. 2: 314, fig. 61e (1963); Killick in Flower. Pl. Afr. 38, fig. 1499 (1967); Palmer & Pitman, Trees S. Afr. 2: 1061 & photo. (1973). Type: South Africa, Umlaas River, *Krauss* 342 (G, iso.!; K, iso.!; M, iso.!; W, iso.!).

Turraea heterophylla sensu Sond. in F.C. 1: 245 (1860); sensu Medley Wood, Natal Plants 3, t. 246 (1902).

Deciduous shrub or small tree up to 9 m tall, sometimes scrambling. *Leaves* ovate to lanceolate. *Inflorescence* a 2−10-flowered fascicle terminating short bracteate shoots behind the current year's growth, rarely in the leaf axils. *Flowers* appearing before or with the new leaves, strongly scented at night. *Petals* 35−60 × 3 mm, pale green, turning pale yellow. *Staminal tube* pure white. *Ovary* 10-locular, puberulous. *Capsule* leathery, becoming woody, obovoid, 15 mm in diameter, deeply sulcate and transversely ribbed. *Seeds* 6 mm long; testa bright orange, contrasting with the whitish, fleshy aril which covers *c*. one third of the seed. *Chromosome number*: 2n = 50 (Zimbabwe, Uganda). Fig. 11: 4.

Widely distributed on the eastern side of Africa from Kenya and Uganda to the eastern Cape. It occurs inside and at the edges of various types of forest, including dune

---

FIG. 12.—**Turraea obtusifolia:** Pictorialized distribution map showing variation in southern Africa of shape and size (all × 0,5) of leaves. The specimens are numbered consecutively from north to south and the degree square from which each specimen was collected is indicated on the map; A, *Smith* 3636 from East London showing variation in leaf shape on a single twig; 1, *Van der Schijff* 3802; 2, *Ihlenfeldt* 2253; 3, *Van der Schijff & Marais* 3671; 4, *Fourie* 2627; 5, *Mogg* 24436; 6, *Hansen* 3296; 7, *Barnard* 387; 8, *Thode* A 1368; 9, *Codd* 722; 10, *De Winter* 8635; 11, *Van Vuuren* 451; 12, *Du Toit* 189; 13, *Pooley* 384; 14, *Reid* 496; 15, *Ward* 4387; 16, *Wells & Edwards* 77; 17, *Mthonti* 1; 18, *Ward* 1844; 19, *Moll* 3110; 20, *Ross & Moll* 2309; 21, *Wood* 5746; 22, *Krauss* 308 from Port Natal (Type of *T. obtusifolia*); 23, *Strey* 8065; 24, *Wylie* s.n.; 25, *Venter* 1035; 26, *De Winter* 8861; 27, *Comins* 1046; 28, *Pegler* 337; 29, *Tyson* s.n. (Kowie); 30, *Tyson* s.n. (Port Alfred); 31, *Jacot Guillarmod* s.n.

2d
2c
2b
2a
2
1e
1c
1
1b
1f
1d
1a
Rosemary Wise.

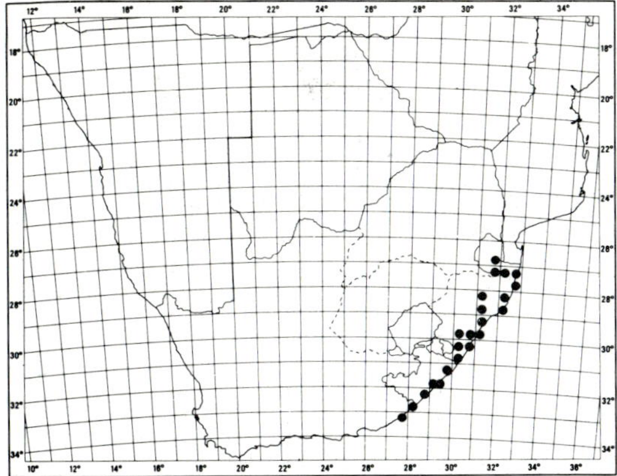

MAP 37.— **Turraea floribunda**

forest. In southern Africa it occupies a narrow coastal belt from Swaziland and northern Zululand to East London, extending up to 175 km inland and from near sea-level to 670 m. Map 37.

Vouchers: *Abbott 402*; *Edwards 1645*; *Schweickerdt 1400*; *Strey 4324*; *Wells 3598*.

The mature leaves of *T. floribunda* are usually entire; two leaves of *Vahrmeyer & Tölken 245*, however, are shallowly 3-lobed.

5. **Turraea pulchella** (*Harms*) *Pennington*

in Blumea 22: 454, fig. 3d (1975). Types: South Africa [Transkei], near Kentani, March 1903, *Pegler 730* (B, holo., presumably destroyed); Kentani Distr., 305 m, fl. fr. 15 Oct. 1904, *Pegler 730* (PRE, neotype of F. White, 1986!; SAM!).

Nurmonia pulchella Harms in Ber. dt. bot. Ges. 35: 80, fig. 1 (1917).

Turraea heterophylla sensu Pegler in Ann. Bolus Herb. 2: 12 (1918).

Suffrutex up to 0,2 m high with tufted herbaceous stems arising from a short (50 mm) branched, woody base, and with a strong, woody, possibly rhizomatous rootstock; branchlets, inflorescence axes and both leaf surfaces sparsely to densely setulose. *Leaves* up to 30×17 mm, obovate-deltate, deeply and bluntly toothed or shallowly lobed in upper half, base cuneate. *Flowers* axillary, solitary, or in lax, 2−3-flowered cymules; peduncle or pedicel 8−12 mm long; bracteoles linear, 2 mm

long. *Calyx* 3−5 mm long, hispidulous, lobes foliaceous, oblong-lanceolate, free almost to the base. *Petals* white, minutely puberulous outside, 7−9×2−3 mm, oblanceolate-oblong, apex rounded, base slightly clawed. *Staminal tube* 5−7 mm long, split at apex for up to one fifth of its length, densely hairy in upper half inside; appendages paired, filiform, as long as anthers; anthers apiculate. *Disc* annular, shallowly lobed. *Ovary* densely hairy, 5-locular, with 2 collateral ovules per locule; style 5 mm long, densely hairy; style-head capitate, less than 1 mm in diameter, with 5 stigmatic lobes at apex. *Capsule* 5-valved and ± 6-seeded, leathery or thinly woody, 4,5×6−7 mm, depressed-globose, shallowly sulcate, densely puberulous. *Seeds* 3×2 mm, very strongly curved, dark brown; aril consisting of a small, whitish, slightly fleshy lobe on each side of the hilum, covering *in toto* about one tenth of the seed. Fig. 13: 1.

Only known from Kentani District in Transkei, where it grows in tall grass between 305 and 610 m. Map 38.

Voucher: *Pegler 730*.

6. **Turraea streyi** *F. White & B. T. Styles*

in Bothalia 16: 153 (1986). Type: South Africa, Natal, St. Michael's-on-Sea, Deppe's Farm, fl., *Strey 6876* (PRE, sheet 1, holo.!; NH!; NU!).

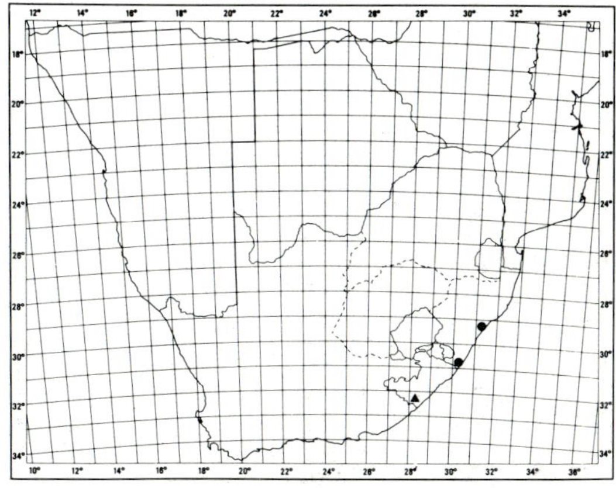

MAP 38.— ▲ **Turraea pulchella**
● **Turraea streyi**

FIG. 13.—1, **Turraea pulchella**: 1, habit, × 0,5; 1a, flower in longitudinal section, × 6,5; 1b, sepal, × 6,5; 1c, petal, × 6,5; 1d, part of staminal tube and anthers, × 6,5; 1e, fruit and leaves, × 0,5; 1f, fruit, × 1,6 (all from *Pegler 730*); 2, **T. streyi**: 2, habit, × 0,5; 2a, flower, in longitudinal section, × 3,2; 2b, anthers and appendages, × 6,5; 2c, fruiting branchlet, × 0,5; 2d, fruit, × 1,6 (all from *Strey 6878*).

1c
1d
1e
1f
1g
1
1a
1b
1b
Rosemary Wise

Suffrutex up to 0,75 m high; branchlets, inflorescence axes and both leaf surfaces sparsely to densely setulose. *Leaves* up to 70 mm long, very variable, trifoliolate, twice-trifoliolate or imparipinnate with 3−5 deeply lobed or trifoliolate divisions; lobes rounded to acute. *Flowers* solitary or in lax 2−3-flowered cymules; pedicel or peduncle slender, up to 35 mm long; bracteoles minute, filiform, ± 1 mm long. *Calyx* 3−5 mm long, hispidulous, lobes foliaceous, oblong-elliptic, acute, free almost to the base. *Petals* white, glabrous, 8−11×3−4 mm, broadly spathulate. *Staminal tube* 7−9 mm long, hairy in upper half inside, not split at apex; appendages paired, filiform, alternating with and much longer than the anthers; anthers long-apiculate, dehiscing by very short slits in lower half. *Ovary* densely hairy, 5-locular, with 2 collateral ovules in each cell; style 7−10 mm long, included or very slightly exserted, hairy in lower half; style-head capitate, less than 1 mm in diameter, with a terminal, 5-lobed coroniform stigma. *Capsule* 5×8 mm, up to 10-seeded, otherwise as in *T. pulchella*. *Chromosome number*: 2n=36 (*Strey* 9288, South Africa). Fig. 13: 2.

Only known from two widely separated localities in the Port Shepstone (3030 CD) and Stanger (2931 AD) degree squares. In grassland and inside and at the edges of scrub forest. Map 38.

Vouchers: *Moll* 5502; *Ross* 1853; *Strey* 9288.

4175   **3. MELIA**

Melia L., Sp. Pl. 1: 384 (1753) & Gen. Pl. edn 5: 182 (1754); Sond. in F.C. 1: 245 (1860); C. DC. in Monogr. Phan. 1: 450 (1878); Harms in Pflanzenfam. edn 2, 19b1: 99 (1940); Exell & Mendonça in C.F.A. 1: 317 (1951); Staner & Gilbert in F.C.B. 7: 172 (1958); White & Styles in F.Z. 2: 315 (1963); R. A. Dyer, Gen. 1: 300 (1975); Pennington & Styles in Blumea 22: 463 (1975); Mabberley in Gdns' Bull., Singapore 37: 49 (1984). Type species: *M. azedarach* L.

Trees or shrubs. *Indumentum* of simple, glandular and tufted-stellate hairs. *Inflorescence* paniculate. *Leaves* 2- or 3-pinnate. *Flowers* bisexual and male on same individual (polygamous). *Petals* 5, free, imbricate. *Staminal tube* narrowly cylindrical; anthers 10, shortly apiculate, alternating with a pair of narrowly deltate appendages. *Disc* annular, crenulate, free from the ovary and staminal tube. *Ovary* with 4−8 locules, each with 2 superposed ovules; style-head capitate with 4−8 short, erect or incurved stigmatic lobes. *Fruit* a 3−8-locular drupe; locules usually 1-seeded.

A small genus with 3 species, 2 of which are indigenous in tropical Africa. *Melia azedarach*, as a wild plant, extends from India to Australia. It is widely planted in the tropics and subtropics.

Melia is the Greek name for the Ash tree of Europe *(Fraxinus excelsior* L., Oleaceae).

Melia azedarach *L.,* Sp. Pl. 1: 384 (1753); Curtis's bot. Mag. 27, fig. 1066 (1807); Sond. in F.C. 1: 246 (1860); C. DC. in Monogr. Phan. 1: 451 (1878); Burtt Davy, Fl. Transv. 2: 486 (1932); O. B. Miller in Jl S. Afr. Bot. 18: 39 (1952); White & Styles in F.Z. 2: 315 (1963); Ross, Fl. Natal 216 (1972); Mabberley in Gdns' Bull., Singapore 37: 55 (1984). Type: Holland, De Hartecamp, cult. *Hort. Cliff.* 161.1 (BM, lecto. of Mabberley, 1984).

Medium-sized rapidly growing short-lived deciduous tree up to 15 m tall, sometimes flowering as a shrub. *Leaves* up to 400 mm long; leaflets up to 55×25 mm, more or less lanceolate, apex acuminate or subacuminate, margin rather deeply crenate or serrate; lower surface sparsely puberulous, glabrescent. *Calyx* 2,5 mm long, densely stellate-puberulous. *Petals* up to 8×3 mm, pale lilac, spathulate. *Staminal tube* up to 7 mm long, dark purple, glabrous outside, hairy inside; appendages *c.* 1 mm long. *Drupe* pale yellow, up to 20×10 mm. *Chromosome number*: 2n=28 (various origins). Fig. 14.

FIG. 14.—**Melia azedarach**: 1, flowering twig, × 0,5; 1a, flower in longitudinal section, × 4,3 (both from *Taylor* 241); 1b, fruits, × 0,5 (after Watt, J. M. & Breyer-Brandwijk, M. G. 1962. *The medicinal and poisonous plants of southern and eastern Africa*, edn 2. Livingstone, Edinburgh & London); 1c−g, stages in germination, × 0,5 (after Troup, R. S. 1921. Meliaceae. *The silviculture of Indian trees* 1: 178−208).

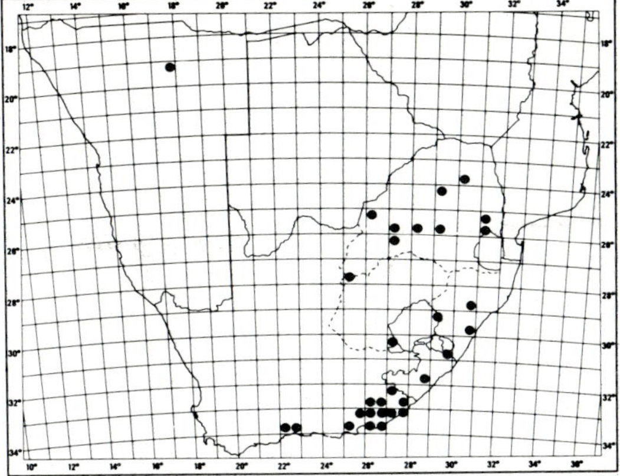

MAP 39.— **Melia azedarach**

Widely planted for ornament in all the warmer parts of southern Africa. In places it is naturalized and has become a pest. As early as 1906 it was established near Barberton, and now, in the Transvaal, it is the most widespread and aggressive of the invasive introduced trees. The cultivar 'Umbraculifera', the Texas Umbrella Tree, a mutant with a flattened crown, is sometimes grown in gardens. Map 39.

Vouchers: *Bayliss* BRI B156; *Edwards* 877; *Onderstall* 746; *Ross* 2379; *Van der Schijff* 3769.

The fruit of *Melia* has long been known to be poisonous to some animals including the pig, and the deaths of children have been reported. Four highly toxic meliatoxins (a group of limonoids) have recently been discovered in the fruit flesh. Fully ripe fruit, however, appears to be much less toxic than fruit just beginning to ripen. The endocarp of the fruits, which are easily pierced at the ends, are used for beads.

4193

4. EKEBERGIA

Ekebergia *Sparrm.* in K. svenska Vetensk-Akad. Handl. 40: 282, t. 9 (1779); Sond. in F.C. 1: 247 (1860); Oliv. in F.T.A. 1: 632 (1868); C. DC. in Monogr. Phan. 1: 641 (1878); Harms in Pflanzenfam. edn 2, 19b1: 119, fig. 29 (1940); Exell & Mendonça in C.F.A. 1: 315 (1951); Keay in F.W.T.A., edn 2, 1: 705 (1958); Staner & Gilbert in F.C.B. 7: 207 (1958); White & Styles in F.Z. 2: 315 (1963); R. A. Dyer, Gen. 1: 300 (1975); Pennington & Styles in Blumea 22: 476 (1975). Type species: *E. capensis* Sparrm.

Trees or shrubs. *Indumentum* of simple hairs. *Leaves* imparipinnate, leaflets entire. *Flowers* pale greenish white, unisexual (plant dioecious), in lax or contracted cymose panicles. *Calyx* saucer-shaped, lobed in upper half. *Petals* 5, imbricate, much longer than the calyx in bud. *Stamens* 10; filaments united for whole or most of length, without appendages. *Disc* annular, fused to base of staminal tube. *Ovary* with 2−5 locules, each with 2 superposed ovules; style-head capitate, with 2−5 small incurved stigmatic lobes. *Fruit* a drupe with (1)2−4(5) pyrenes. *Seeds* mostly 1 per locule, exarillate.

Character not applicable in southern Africa: one species is a geoxylic suffrutex.

A small genus with 4 species confined to the African mainland.

The name *Ekebergia* is in honour of Charles Gustavus Ekeberg, captain of the Swedish East Indiaman in which Sparrman sailed to China.

Leaf rhachis unwinged or very narrowly winged .... 1. *E. capensis*
Leaf rhachis broadly winged .... 2. *E. pterophylla*

1. Ekebergia capensis *Sparrm.* in K. Svenska Vetensk-Akad. Handl. 40: 282, fig. 9 (1779); Thunb., F.C. edn Schult. 542 (1823); Sond. in F.C. 1: 247 (1860) excluding *Ecklon & Zeyher* 425; Pappe, Silv. Cap. 6 (1862); C. DC. in Monogr. Phan. 1: 641 (1878); Burtt Davy, Fl. Transv. 2: 487 (1932); White & Styles in F.Z. 2: 316, fig. 62 (1963); Ross, Fl. Natal 216 (1972); Van Wyk, Trees Kruger Nat. Park, 1: 274 & photos (1972); Palmer & Pitman, Trees S. Afr. 2: 1065 & photos (1973). Type: South Africa, *Sparrman* s.n. (S, holo.!).

Trichilia ekebergia E. Mey. ex Sond. in F.C. 1: 246 (1860); E. Mey. ex Drège, Zwei Pfl. Doc. 227 (1843), nom. nud.; Pappe, Silv. Cap. 5 (1862). Syntypes: South Africa, Tsitsikamma ('Sitzekamma'), *Drège* s.n. (TCD; S); Port Natal, *Krauss* (TCD; S).

Ekebergia meyeri Presl ex C. DC. in Monogr. Phan. 1: 642 (1878); Presl, Bot. Bemerk. 25 (1844), nom. nud. (as *Eckebergia*); Burtt Davy, Fl. Transv. 2: 487 (1932). Type:

FIG. 15.—1, **Ekebergia capensis:** 1, flowering twig, × 0,5 (from *Bos* 9720); 1a, female flower in longitudinal section, × 10,6 (from *White* 8146); 1b, infructescence, × 0,5; 1c, fruit in transverse section, × 1,1 (both from *De Wilde* 5992); 2, **E. pterophylla:** 2, fruiting twig, × 0,5 (from *White* 10308); 2a, inflorescences, × 0,5 (from *Compton* s.n.).

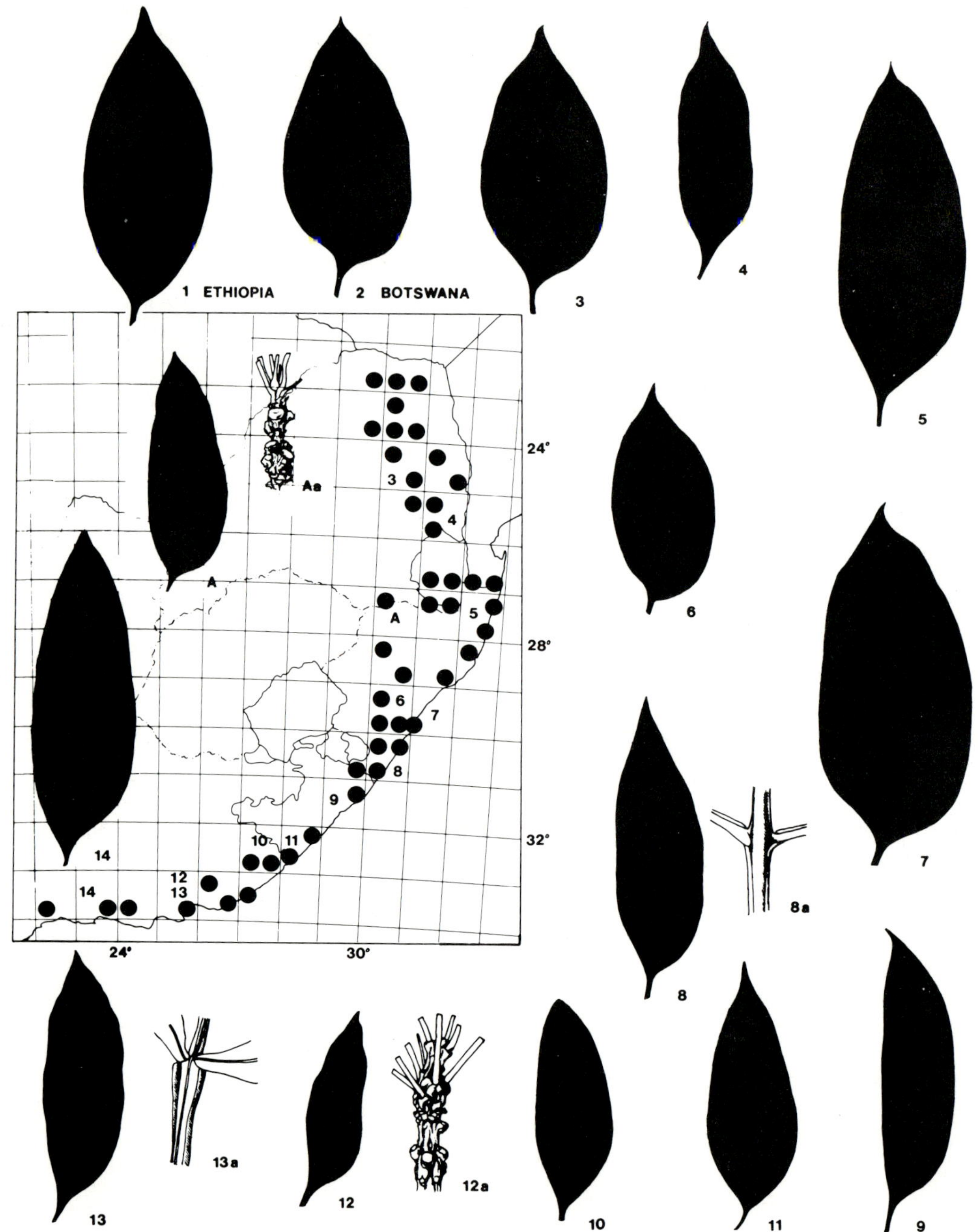

FIG. 16.—**Ekebergia capensis:** Pictorialized distribution map showing variation in southern Africa of: leaflet shape and size, leaf rhachis, and leaf scars on branchlets. Specimens are numbered consecutively from north to south. For leaflets 3–14 its number is also shown on the map inside the degree square from which the specimen it came from was collected. 1, *Bos* 9720; 2, *Smith* 1471 from Dikgathong, SE 1922; 3, *Van Greuning* 502; 4, *De Souza* 81; 5, *Moll* 4927; 6, *Moll* 1724; 7, *Ward* 5848; 8, *White* 10527; 9, *Strey* 10123; 10, *Flanagan* 818; 11, *Pegler* 849; 12 and 12a, *Transvaal Museum*; 13 and 13a, *Olivier* 1101; 14, *Marloth* 7433. Leaflets and branchlets, × 0,5; leaf rhachis, × 2. A, *Devenish* 1039 from northern Natal showing remarkable similarity to the Cape variants.

South Africa, 'Cape Peninsula', *Drège* s.n. (B, holo., presumably destroyed).

Evergreen or semi-evergreen tree up to 20 m tall, often with a wide, umbrageous crown. *Leaflets* (7)9—13, lanceolate to oblong-lanceolate, up to 145×60 mm, tapering to an acuminate, subacuminate or, rarely, acute apex, base asymmetric; lower surface glabrous to tomentose. *Inflorescence* 70—250 mm long. *Flowers* white or pinkish white. *Calyx* 2 mm long, sparsely to densely puberulous. *Petals* 4—5 mm long, elliptic-oblong, densely puberulous on both surfaces. *Staminal tube* 2 mm long, puberulous outside, densely bearded at the throat inside. *Ovary* 15×2—2,5 mm, densely setulose; style 0,5 mm long. *Drupe* deep red, 18×15 mm, with (1)2—4(5) pyrenes. *Chromosome number*: 2n=46 (*Vosa & Burras* 452, South Africa). Figs 15: 1 & 16.

Widespread in tropical Africa from Senegal to Ethiopia and southwards to Zambia, Botswana and South Africa. In various types of forest and scrub forest from near sea-level to 1 675 m. Map (see fig. 16).

Vouchers: *Codd* 6690; *Codd & Dyer* 4471; *Kluge* 580; *Mauve* 5254; *Olivier* 1101.

The Dog plum or Essenhout, like most widespread species, is very variable. Most individuals from the eastern Cape have very prominent leaf scars, a narrowly winged rhachis and small, narrow, crowded leaflets. In tropical Africa and Natal the predominant form has a wingless rhachis and larger, broader, more widely spaced leaflets. These two variants, however, have been confused since before 1860. There are also too many intermediates and the overall pattern is too diffuse to justify the formal recognition of named taxa. The Cape variant is often cultivated as a street tree and in gardens. *E. capensis* is sometimes confused with *Harpephyllum caffrum* Bernh. (Anacardiaceae), the leaves of which are superficially similar.

2. Ekebergia pterophylla *(C. DC.) Hofmeyr* in J. Bot., Lond. 63: 57 (1925); Ross, Fl. Natal 216 (1972); Palmer & Pitman, Trees, S. Afr. 2: 1068 & photos (1973). Type: South Africa, near 'Gwenberg', *Medley* ('Imley') *Wood* 1022 (G, holo.; K!).

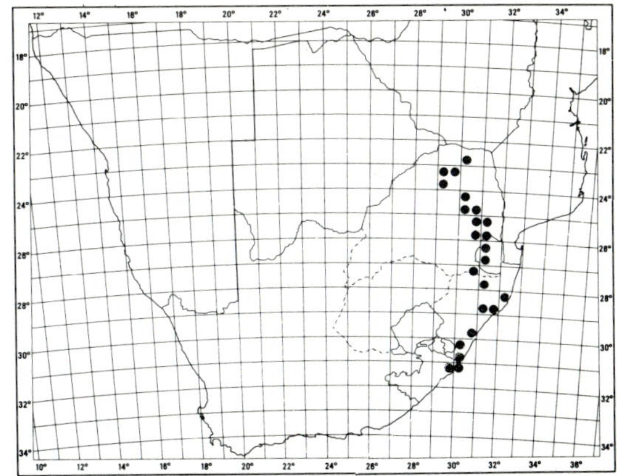

MAP 40.— **Ekebergia pterophylla**

Trichilia pterophylla C. DC. in Bull. Herb. Boissier 2: 581 (1894); Burtt Davy, Fl. Transv. 2: 487 (1932).

Trichilia alata N.E.Br. in Kew Bull. 1896: 160 (1896); Wood & Evans, Natal Plants 2, fig. 209 (1902). Syntypes: Transvaal, Barberton, Upper Moodies, *Galpin* 1083 (K!; PRE!); Natal, near Pinetown, *Medley Wood* 1043 (K!; PRE!; etc.).

Very similar to *Ekebergia capensis* (no. 1) in its basic morphology though very dissimilar in appearance, differing as follows: Shrub or small tree, usually less than 6 m tall. *Leaflets* 3—5(7), coriaceous, elliptic, obovate-elliptic or oblanceolate-elliptic, apex mostly rounded or retuse, margin recurved, up to 50×20 mm. *Inflorescence* up to 60 mm long. *Drupe* up to 10×10 mm. *Chromosome number*: 2n=50 (*White* 10523, South Africa). Fig. 15: 2.

Only known from South Africa where it extends from the northern Transvaal to Pondoland. Usually in semi-evergreen bushland and scrub forest and at forest edges, especially in rocky places. Mostly between 500 and 1 700 m. Map 40.

Vouchers: *Acocks* 12860; *Compton* 25926; *Gerstner* 6027a; *Strey* 10627; *Van Wyk & Venter* 1294.

The Rock Essenhout is an unmistakable species though it has been confused with *Schrebera alata* (Hochst.) Welw. (Oleaceae).

4195   5. TRICHILIA

Trichilia *P. Browne*, Hist. Jam. 278 (1756) nom. conserv.; Sond. in F.C. 1: 246 (1860); Oliv. in F.T.A. 1: 333 (1868); Harms in Pflanzenfam. edn 2, 19b1: 104 (1940); Exell & Mendonça in C.F.A. 1: 312 (1951); Keay in F.W.T.A. edn 2, 1: 703 (1958); Staner & Gilbert in F.C.B. 7: 157 (1958); White & Styles in F.Z. 2: 297 (1963); De Wilde, Rev. spec. *Trichilia* Afr. (1968); R. A. Dyer, Gen. 1: 300 (1975); Pennington & Styles in Blumea 22: 467 (1975). Type species: *T. hirta* L.

Trees. *Indumentum* of simple hairs. *Leaves* imparipinnate; leaflets entire. *Flowers* unisexual (plants apparently dioecious), in cymes or cymose panicles. *Sepals* 5. *Petals* 5, imbricate, much

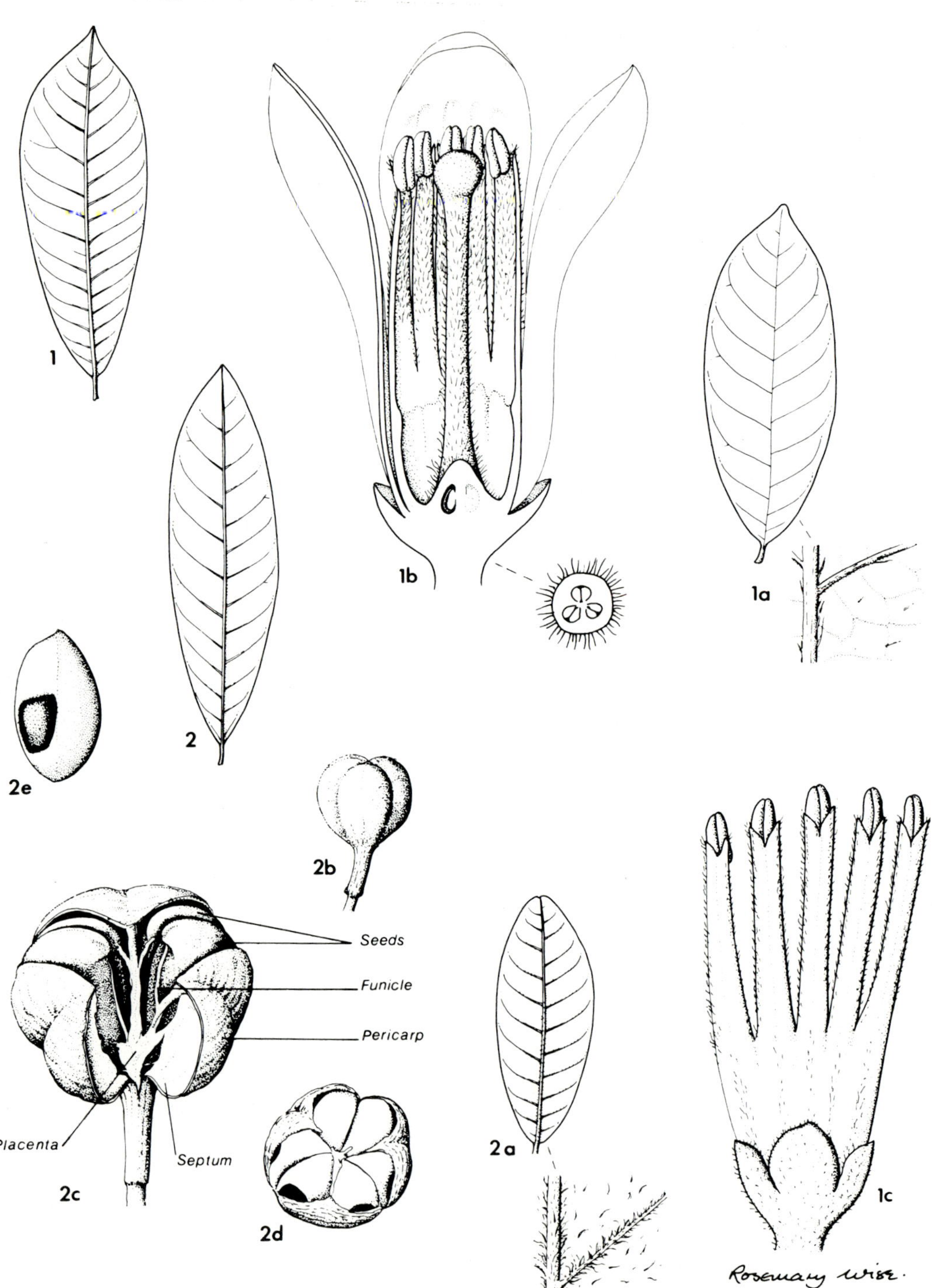
Seeds
Funicle
Pericarp
Placenta
Septum
Rosemary Wise.

longer than the calyx in bud. *Stamens* 10; filaments united in lower half and with a pair of deltate appendages at apex. *Disc* fused to base of staminal tube. *Ovary* small; locules with 2 collateral ovules. *Fruit* a loculicidal capsule with 3 leathery valves and up to 6 seeds. *Seeds* large, almost completely covered by a bright red aril.

Characters not applicable in southern Africa: suffruticose habit, stellate hairs, unifoliolate or trifoliolate leaves, completely united filaments without appendages, free cup-shaped disc, bilocular ovary.

A mainly tropical genus of 85 species, most numerous in the New World, with only 15 species in Africa and 2 in Asia.

Trichilia natalensis Sond. is not meliaceous and *T. umbellata* C. DC. ex Medley Wood and ex Bews is a *nomen nudum* (see De Wilde, 1968).

Derived from a Greek word meaning 'in three parts' referring to the fruits which are 3-lobed in some species.

Until recently the two southern African species have been repeatedly confused. They are ecologically and, to a large extent, geographically distinct and the trained eye has no difficulty in distinguishing them. They differ in several features of the leaflets, fruits and infructescence, but some characters show slight overlap. The most reliable character concerns the stipe of the fruit and this is the only one used in the key below. There should, however, be no difficulty in identifying flowering and sterile material by referring to the descriptions and illustrations.

Capsule without a stipe or with an indistinct stipe up to 3 mm long.. 1. *T. dregeana*

Capsule sharply differentiated from a 5 − 10 mm long stipe... 2. *T. emetica*

1. **Trichilia dregeana** *Sond.* in F.C. 1: 246 (1860); White & Styles in F.Z. 2: 298 (1963); De Wilde, Rev. spec. *Trichilia* Afr. 28, fig. 3, map 3 (1968); Ross, Fl. Natal 216 (1972); Palmer & Pitman, Trees S. Afr. 2: 1069 & photos (1973). Type: South Africa, Port Natal (Durban), *Gueinzius* s.n. (TCD, lecto. of De Wilde, 1968; G!; K!; PRE!; S; SAM!).

Trichilia dregeana var. *oblonga* Harv. ex Sond. in F.C. 1: 246 (1860); Harv., Thes. Cap. 1: 49 fig. 76 ('1859', probably 1860, see below). Type: South Africa, Port Natal [Durban], *Sanderson* s.n. (TCD, holo.; K!).

Trichilia dregei E. Mey. ex C. DC. in Monogr. Phan. 1: 657 (1878); E. Mey. ex Drège, Zwei Pfl. Doc. 227 (1843), *nom. nud.* Type: *Drège* 39 (P, holo.!).

Trichilia dregei var. *oblonga* (Harv. ex Sond.) C. DC. in Monogr. Phan. 658 (1878), *nom. illegit.*

Trichilia emetica sensu Burtt Davy, Fl. Transv. 2: 487 (1932), as to syn. *T. dregeana* only.

Evergreen tree up to 20 m tall; bole up to 1,8 m diam., tall and straight when growing inside forest, but branched at 2 m in open; crown dense, very wide-spreading in open. *Leaves* up to 260 mm long. *Leaflets* 7 − 11, up to 210 × 85 mm; apex of lateral leaflets shortly and acutely acuminate or bluntly subacuminate (with the tip itself often shallowly notched), nearly always showing a hollow curve; lower surface glabrous or with a few short strigulose hairs, usually drying dark brown; lateral nerves in 7 − 12, usually widely spaced pairs. *Inflorescence* sometimes many-flowered in male, al- ways few-flowered and usually lax in female. *Petals* usually 13 − 22 mm long. *Staminal tube* usually 10 − 16 mm long. *Fruit* (ripe but un- opened) 30 − 50 mm in diam., occurring 1 − 3 together in the leaf axils. *Chromosome number*: 2n = *c*.360 (Ethiopia). Fig. 17: 1.

Widespread in tropical Africa, but with some major disjunctions; mainly in upland areas from Guinea to Camer- oun and from Ethiopia to South Africa; also in Angola. In South Africa in evergreen forest as far south as Komga District in the eastern Cape; mostly in regions of higher rainfall than *T. emetica;* from near sea-level to 1 250 m. Map 41.

Vouchers: *De Winter* 8850; *Moll* 3226; *Ross* 1270; *Scheepers* 1251; *Strey* 7122.

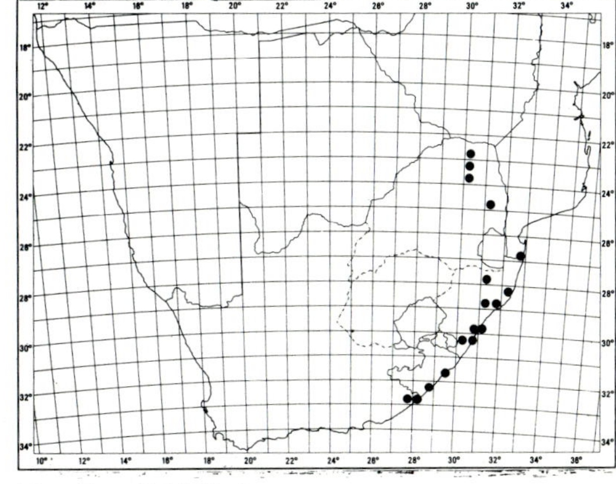

MAP 41.— **Trichilia dregeana**

FIG. 17.—1, **Trichilia dregeana:** 1, leaflet, × 0,5 (from *Thorns* 5348); 1a, leaflet, × 0,5 and part of lower leaf surface, × 2,1 (both from *Strey* 5340); 1b, flower in longitudinal section and ovary in transverse section (both × 3,2 and from *Styles* 333); 1c, calyx and androecium, × 3,2 (from *Styles* 333); 2, **T. emetica:** 2, leaflet, × 0,5 (from *Liengme* 577); 2a, leaflet, × 0,5 and part of lower leaf surface, × 2,1 (both from *Campton* 31147); 2b, unopened capsule, × 0,5; 2c, capsule just beginning to dehisce, × 1,6; 2d, dehisced capsule viewed from above, × 1,1; 2e, seed, × 1,6 (all from *Bond* s.n.).

FIG. 18.—**Pseudobersama mossambicensis:** 1, fruiting twig, × 0,5 (from *Boococh* 16; 1a, male flower in longitudinal section, × 7,5 (after Pennington & Styles); 1b, capsule, × 0,5; 1c, seed, × 2,1 (both from *Moggridge* 430).

2. Trichilia emetica *Vahl*, Symb. Bot. 1:
31 (1790); Burtt Davy, Fl. Transv. 2: 487
(1932), excl. syn. *T. dregeana*; O. B. Miller in
Jl S. Afr. Bot. 18: 40 (1952); White & Styles in
F.Z. 2: 299, fig. 58b (1963); De Wilde, Rev.
spec. *Trichilia* Afr. 50, fig. 4, map 4 (1968);
Ross, Fl. Natal 216 (1972); Van Wyk, Trees
Kruger Nat. Park 1: 277 & photos (1972);
Palmer & Pitman, Trees S. Afr. 2: 1071 & photos (1973). Type: Yemen, Forsskål 478 (C,
holo.; BM!).

Trichilia roka Chiov. in Bull. Soc. Bot. Ital. 1923: 115
(1923); F. White, For. Fl. N. Rhod. 181 (1952), *nom.
illegit.* Type: as above.

Evergreen or semi-evergreen tree up to 18
m tall, but in southern Africa usually smaller;
crown very dense, widespreading in open.
Leaves up to 280 mm long. *Leaflets* 9−11, up
to 150 × 50 mm; apex of lateral leaflets
rounded, emarginate or broadly acute, without
a hollow curve; lower surface sparsely to
densely puberulous or tomentellous with short
weak curly or flexuose hairs, usually drying
olive-green or pale yellow-brown; lateral nerves
usually in 11−18 closely set pairs. *Inflorescence* usually condensed and many-flowered.
Petals usually 9−14 mm long. *Staminal tube*
usually 8−11 mm long. *Fruit* (ripe but unopened) 18−25 mm in diameter, usually
crowded at the ends of the branchlets. *Chromo-*

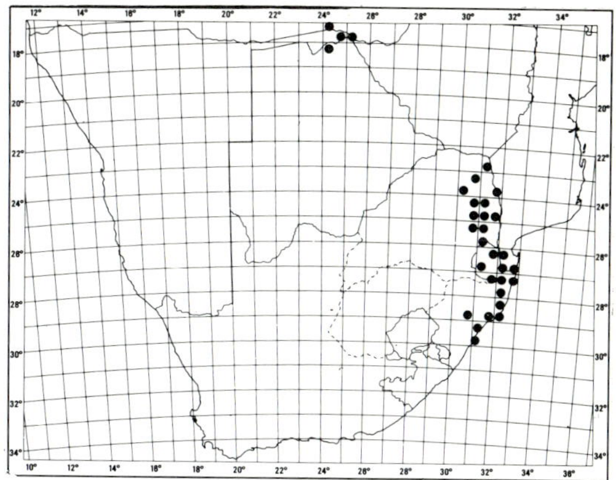

MAP 42.— **Trichilia emetica**

some number: 2n = 50 (*Vosa & Burras* 453;
White 10415, South Africa). Fig. 17: 2.

Widespread in tropical Africa from Senegal to the Red
Sea and the Yemen, and through East and Central Africa to
the Caprivi Strip, Botswana and Natal. In the Caprivi Strip
and Botswana it is a characteristic member of riparian forest
fringing the Zambezi and Chobe Rivers. In southern Africa
it mostly occurs in drier areas than *T. dregeana* (no. 1);
usually in riparian forest but also locally in woodland and
wooded grassland on and in scrub forest in rocky places. In
Natal it occurs in coastal forest from Mtunzini northwards,
whereas *T. dregeana* occupies that habitat further south.
Map 42.

Vouchers: *Acocks* 12974; *Edwards* 3326; *Pole Evans*
4201; *Strey* 3243; *Von Breitenbach* 1217.

4195a

6. PSEUDOBERSAMA

Pseudobersama *Verdc.* in J. Linn. Soc., Bot. 55: 504 (1956); White & Styles in F.Z. 2: 304
(1963); R. A. Dyer, Gen. 1: 301 (1975); Pennington & Styles in Blumea 22: 469 (1975). Type
species: *P. mossambicensis* (Sim) Verdc.

Trees with pinnate leaves. *Indumentum* of simple hairs. *Flowers* unisexual (plant dioecious).
Petals 5, free, imbricate. *Filaments* united for about half of length, without appendages; anthers
11−12, hairy. *Disc* partly fused to base of ovary. *Ovary* with 5 locules each with 2 collateral
ovules; style-head scarcely broader than style, obscurely 5-lobed. *Capsule* with 2 seeds per locule.
Seeds very small in relation to size of capsule, partly surrounded by a small red, fleshy aril.

One species on the eastern side of Africa.

The generic name means 'false *Bersama*'.

Pseudobersama mossambicensis *(Sim)*
Verdc. in J. Linn. Soc., Bot. 55: 504, fig. 1
(1956); White & Styles in F.Z. 2: 305, fig. 60
(1963); Ross, Fl. Natal 216 (1972). Type: Mozambique, without locality, *Sim* 5204 [not
located; fig. 23 in Sim, For. Fl. P.E. Afr.
(1909), lectotype of White (1986)].

Bersama mossambicensis Sim, tom. cit. 34, fig. 23
(1909).

Evergreen tree up to 10 m tall, usually
smaller. *Leaves* up to 300 mm long. *Leaflets*
9−17, elliptic or oblong-elliptic, apex shortly
and bluntly acuminate; lower surface with tufts
of hairs in the nerve axils and with lax, promi-

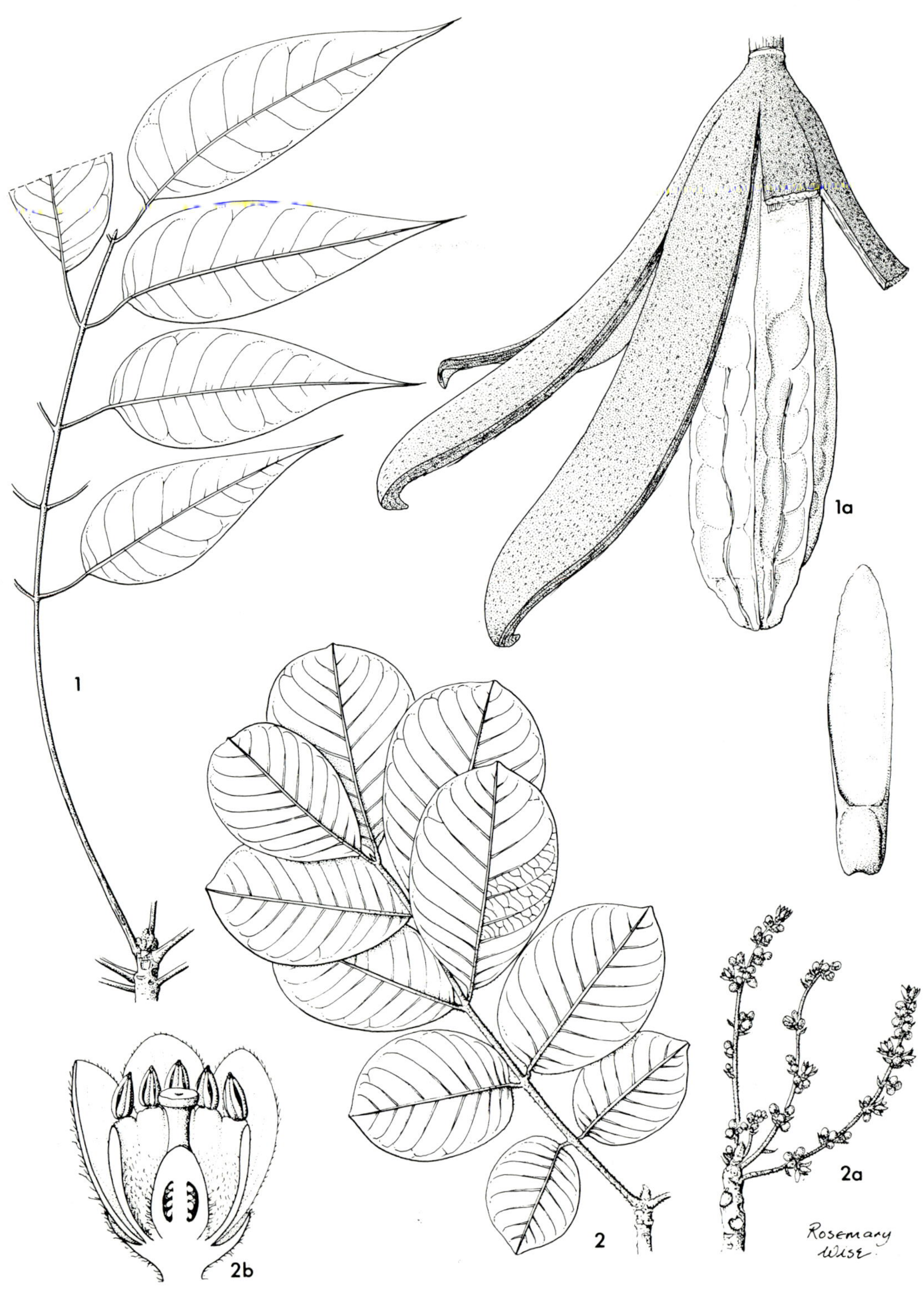
Rosemary
Wise

nent, reticulate venation. *Flowers* white, 3—12 in a sparsely branched to subcapitate, cymose inflorescence. *Petals* 4,5—6 mm long, glabrous. *Filaments* 3,5 mm long. *Capsule* 30—50 mm in diameter, red; appendages up to 7 mm long. *Seeds* 7 × 5 mm, purple-black, aril orange, extending from the adaxial part of seed, to form a cushion at the apex and base. *Chromosome number*: 2n = 46 (Kenya). Fig 18.

P. mossambicensis has a scattered distribution in east African coastal forests from Kenya to Natal. Map 43.

Vouchers: *Tinley* 321; 456; *Ward* 3231.

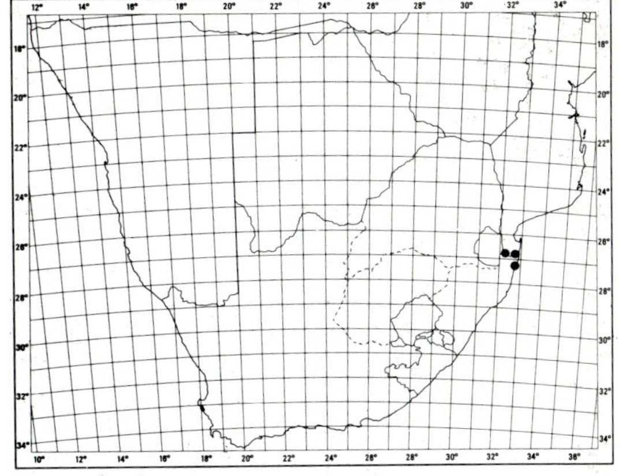

MAP 43.— **Pseudobersama mossambicensis**

4163   7. ENTANDROPHRAGMA

Entandrophragma *C. DC.* in Bull. Herb. Boissier 2: 582, fig. 12 (1894); Harms in Pflanzenfam. edn 2, 19b1: 55 (1940); Exell & Mendonça in C.F.A. 1: 308 (1951); White & Styles in F.Z. 2: 289 (1963); M. Friedrich in F.S.W.A. 71: 2 (1968); R. A. Dyer, Gen. 1: 298 (1975); Pennington & Styles in Blumea 22: 518, t. 17, i, j (1975). Type species: *E. angolense* (Welw.) C. DC.

Wulfhorstia C. DC. in Mém. Herb. Boissier 10: 77 (1900).

Trees, usually large. *Leaves* paripinnate, leaflets entire. *Flowers* unisexual (tree monoecious), in contracted panicles. *Petals* 5, free, contorted. *Staminal tube* cup-shaped, margin entire or almost so; anthers 10; appendages absent. *Disc* in the form of 10 distinct or indistinct short ridges which connect the ovary to the base of the staminal tube. *Ovary* 5-locular, each locule with 4—12 pendulous anatropous ovules in 2 rows; style short; style-head discoid with 5 radiating stigmatic lines. *Capsule* pendulous, woody, septifragal, cigar-shaped; valves separating from apex, persistent at base; columella massive, softly woody, 5-angled, indented with imprints of the seeds. *Seeds* with a terminal wing, 3—5 per locule.

Characters not applicable in southern Africa: the precise shape of the staminal tube and disc and the way the fruit opens are different in some other species.

A genus with 11 species, confined to the African mainland.

The Greek name *Entandrophragma* refers to the 10 partitions between the ovary and the staminal tube.

Leaflets filiform-acuminate at apex ....1. *E. caudatum*
Leaflets shortly cuspidate to emarginate at apex...2. *E. spicatum*

1. **Entandrophragma caudatum** *(Sprague) Sprague* in Kew Bull. 1910: 180 (1910) & in Hooker's Icon. Pl. 31: fig. 3032 (1915); Burtt Davy, Fl. Transv. 2: 487 (1932); Brem. & Oberm. in Ann. Transv. Mus. 16: 420 (1935); O. B. Miller in Jl S. Afr. Bot. 18: 39 (1952); White & Styles in F.Z. 2: 290, fig. 55b (1963); Ross, Fl. Natal 216 (1972); Van Wyk, Trees Kruger Nat. Park 1: 271 & photos (1972); Palmer & Pitman, Trees S. Afr. 2: 1055 & photos (1973). Type: Transvaal, Blouwberg, 1 190 m, *Bailey* in Transv. Dept. Agric. Herb. 2926 (K, holo.!).

Pseudocedrela caudata Sprague in Kew Bull. 1908: 163 (1908).

Large deciduous treè 10—20 m tall; bole up to 1,5 m diam., long and straight or branched at 2 m. *Bark* thick and rough, flaking in large

---

FIG. 19.—1, **Entandrophragma caudatum:** 1, leaf, × 0,5 (from *Smith* 2053); 1a, capsule and seed, × 0,5 (from *Burtt Davy* 20573); 2, **E. spicatum:** 2, leaf × 0,5 (from *Teixeira* 2414); 2a, inflorescence, × 0,5; 2b, flower in longitudinal section, × 8 (both from *Gossweiler* 13375).

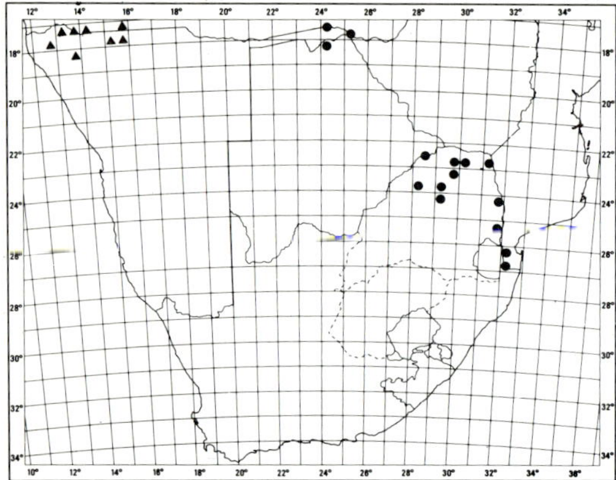

MAP 44.— • **Entandrophragma caudatum**
          ▲ **Entandrophragma spicatum**

irregular scales to expose buff undersurface. *Leaves* up to 250 mm long. *Leaflets* in 5—8 pairs, up to 110 × 35 mm, ovate to lanceolate, gradually tapering from near the base to a narrowly acuminate apex with a filiform tip, base of distal leaflets very asymmetric, base of proximal ones less asymmetric; lower surface glabrous or almost so, venation indistinct. *Inflorescence* up to 200 × 20 mm, in leaf axils and towards base of current year's growth. *Flowers* pale green. *Petals* 5—6 mm long, densely puberulous outside. *Staminal tube* 3—4 mm long, glabrous outside, densely puberulous inside. *Capsule* 150—200 × 50 mm, cigar-shaped; valves brownish black, closely lenticellate, apex thickened and recurved; columella with shallow seed impressions. *Seeds* including the wing 90—100 × 20 mm. *Chromosome number*: 2n = 72 (Zimbabwe). Fig. 19: 1.

From Zambia and Malawi southwards to Swaziland and northern Zululand. A conspicuous, though rather rare, emergent tree in various types of bushland, thicket and forest on sandy soils or in rocky places, including *Baikiaea plurijuga* forest on Kalahari Sand in Botswana and the Caprivi Strip, and 'mahemane' thicket at the southern end of its range in Zululand. Map 44.

Vouchers: *De Winter* 9241A; *Hancock* 30; *Smith* 2053; *Vahrmeijer* 1542; *Von Breitenbach* 1225.

The wood of Mountain mahogany can be used for furniture and cabinet making but supplies are too limited for it to be of much economic importance.

2. Entandrophragma spicatum *(C. DC.)* Sprague in Hooker's Icon. Pl. 31: in nota sub fig. 3023 (1915); Friedr. in F.S.W.A. 71: 2 (1968). Type: S.W.A./Namibia, Amboland,

Uukuanjama, Omupanda, *Wulfhorst* s.n. (Z, holo.).

Wulfhorstia spicata C. DC. in Mém. Herb. Boissier 10: 78 (1900).

Wulfhorstia spicata var. *viridiflora* Schinz in Bull. Herb. Boissier, sér. 2,2: 1000 (1902). Type: S.W.A./Namibia, Omupanda, *Rautanen* 323 (Z, holo.).

Wulfhorstia ekebergioides Harms in Warb., Kunene-Samb. Exped. 271 (1903). *Entandrophragma ekebergioides* (Harms) Sprague in Hooker's Icon. Pl. 31: in nota sub. fig. 3023 (1915); Exell & Mendonça, C.F.A. 1: 309 (1951). Type: Angola, Huila, Vila Pereira d'Eca, *Baum* 88 (B, holo., presumably destroyed; BM!; COI!; K!).

Large deciduous tree 9—18 m tall. *Bark* rough, flaking. *Leaflets* in 3—7 pairs, up to 110 × 70 mm, broadly oblong or obovate-oblong; lower surface densely puberulous, venation forming a conspicuous network. *Inflorescence* densely hairy, precocious or with the leaves. *Capsule* with greyish, buff or pale brown valves; columella with deep seed impressions. *Chromosome number*: 2n = 72 (Angola). Fig. 19: 2.

The Woodland mahogany is only known from the south of Angola and the north of S.W.A./Namibia, where it is an emergent tree from bushland and woodland, mostly on deep sandy soil. Map 44.

Vouchers: *De Winter* 3634; *De Winter & Leistner* 5814; *Rodin* 8914.

Introduced species

In addition to *Melia azedarach* (p. 49), voucher specimens have been seen for the following species which are planted. Six other species are included in the *National list of introduced trees* (Von Breitenbach, 1984).

Aphanamixis polystachya *(Wall.) R. N. Parker* (syn. *A. rohituka* Pierre). Native of tropical Asia. Tree 12 m tall. Leaves pinnate. Leaflets very asymmetric, entire; venation prominently reticulate. Inflorescence spicate or paniculate. Female and hermaphrodite flowers sessile. Male flowers sometimes stalked. Fruit a 2—3(4)-valved loculicidal capsule. Seeds with a bright red aril. In parks in Durban; said to be 'wild' in forest at Hluhluwe (Palmer & Pitman, 1972), but almost certainly planted; more information needed). Vouchers: *Mills* 23; *White* 14127.

Azadirachta indica *A. Juss.* (syn. *Melia azadirachta* L.). Neem or Nim Tree. Native of India and Burma. Tree. Leaves pinnate. Leaflets very asymmetric, coarsely serrate. Inflorescence paniculate. Fruit a drupe. Planted at Mtunzini. Voucher: *Haigh* CPF 2448.

Cedrela odorata *L.* Central American Cedar, Midde-Amerikaanse Seder. Native of tropical America. Tree up to 30 m high. Leaves pinnate. Leaflets entire, foetid when crushed, oblong-lanceolate; proximal lateral nerves with axillary pockets on lower surface. Inflorescence paniculate. Fruit a robust, woody capsule usually 30—40 mm long. Seeds winged. Planted for timber and ornament. Vouchers: *Wells & Edwards* 18; *NH* 30505.

Khaya nyasica *Stapf ex Bak. f.* East African Mahogany; Oos-Afrikaanse Mahonie. Native of tropical Africa. Tree up to 25 m or more high. Leaves pinnate. Leaflets entire. Inflorescence paniculate. Fruit an almost spherical woody capsule 40—60 mm in diameter. Seeds transversely ellipsoid, narrowly winged all round margin. Successfully planted for shade and ornament, and for timber. Voucher: *Van Rensburg* 411.

Swietenia mahagoni *(L.) Jacq.* Spanish Mahogany; Spaanse Mahonie. Native of Florida and the West Indies. Tree up to 20 m tall. Leaves pinnate. Leaflets distinctly petiolulate, entire, not foetid, lanceolate, very asymmetric, apex subacuminate. Inflorescence paniculate. Fruit a robust woody capsule 60—100 mm long. Seeds winged. Planted in Durban. Voucher: *Moll* 3621.

Toona ciliata *M. J. Roem.* (syn. *Cedrela toona* Roxb. ex Rottl. & Willd.). Toon Tree; Toonboom. Native of tropical Asia. Tree up to 25 m or more high. Leaves pinnate. Leaflets entire, not foetid, lanceolate or lanceolate-elliptic, without axillary pockets beneath. Inflorescence paniculate. Fruit a delicate woody capsule ± 20 mm long. Seeds winged. Planted for timber and ornament. Vouchers: *Rogers* 14829; *Wyman* 17.

MALPIGHIACEAE nom. cons.

by Various Authors

Scandent or erect shrubs or lianes, rarely small trees; with medifixed unicellular hairs. *Leaves* opposite or alternate, rarely ternate, simple, entire, with sessile or stipitate glands on petiole or base of lamina; stipules present or absent. *Inflorescence* of axillary umbels, axillary corymbs or terminal racemes. *Flowers* bisexual, with 2 bracts on pedicel, slightly zygomorphic. *Sepals* 5, free, may have 1 or 2 sessile glands on outer surface. *Petals* 5, free, spreading, unguiculate, with fimbriate or lacerate margins. *Stamens* 10, free or filaments shortly connate at base; filaments shorter than anthers, may be unequal; anthers opening longitudinally or by an apical pore-like slit. *Ovary* superior, 3 (4)-locular, one locule may abort; styles 2−3 (4), free to the base; placentation axile, each locule with a single pendulous ovule. *Fruit* breaking up into 2−3 (4) winged samaras; each with a dorsal or lateral wing.

Characters not applicable to species in our area: May have stinging hairs, hairs may be branched. *Leaves* with jointed petioles. *Flowers* may be cleistogamous. *Stamens*—some reduced to staminodes; anthers with an enlarged connective. *Ovary* may be 5-locular; styles rarely connate. *Fruit* may be a schizocarp, capsule, berry or drupe.

Genera about 60; species about 850; cosmopolitan in tropical and subtropical areas with most species in the American tropics and subtropics. Three genera and 4 species occur in southern Africa.

1a Leaves alternate; styles 2 ....3. **Acridocarpus**

1b Leaves opposite or ternate; styles 3:

    2a Flowers yellow or orange; fruit with obliquely oblong dorsal wing . 2. **Sphedamnocarpus**

    2b Flowers pink, fading whitish; fruit with a saucer-like lateral wing ....1. **Triaspis**

4206
1. TRIASPIS

by K. L. Immelman

Triaspis *Burch.*, Trav. 2: 280, 290 (Jan.−June 1824); Sond. in F.C. 1: 232 (1859); Benth. & Hook. f., Gen. Pl. 1: 259 (1862); R. A. Dyer, Gen. 1: 302 (1975). Type species: *T. hypericoides* (DC.) Burch.

Scandent shrubs with twining stems, sometimes erect shrubs. *Leaves* opposite, rarely ternate, simple, entire, with or without glands on lamina; stipules absent. *Flowers* in axillary corymbose racemes; bracteate on pedicels, bisexual, slightly zygomorphic. *Sepals* 5, free, glabrous or pilose on abaxial surface. *Petals* 5, free, spreading, concave, shortly unguiculate, margins fimbriated; one petal long-fimbriate along whole margin, one petal long-fimbriate at base only, three petals long-fimbriate along one side and at base of the other, short-fimbriate near top. *Stamens* 10, free; filaments often slightly unequal. *Ovary* superior, 3-locular, with single pendulous ovule in each loculus; styles 3, free to base, curving. *Fruit* breaking up into (2−)3 one-seeded carpels, each with a shield- or saucer-shaped wing on margin.

Species about 12, all African, with two in southern Africa. These occur in S.W.A./Namibia, Botswana, northern Cape and Transvaal.

Wing on carpel 1,2−2,0 times as long as wide, 10−25 mm wide; leaf 0,7−1,6 times longer than wide; petiole (2−)4−16(−20) mm long..1. *T. glaucophylla*

Wing on carpel 0,8−1,3 times as long as wide, 25−40 mm wide; leaf (1−)1,2−3,0(−6,5) times longer than wide; petiole 1−5(−7) mm long...2. *T. hypericoides*

    1. **Triaspis glaucophylla** *Engl.* in Bot. Jb. 36: 249 (May, 1905); Niedenzu in Pflanzenreich 4, 141 (Heft 91): 48 (1928). Type: Transvaal, N.W. of Lydenburg, by big waterfall, *Wilms* 144 (B?†).

T. leendertziae Burtt Davy, Fl. Transv. 1: 51 (1926); 2: 286 (1932). Type: Transvaal, Potgietersrus, *Leendertz* 1962 (K, holo.!; BOL!).

Scandent shrub, up to 2(−5) m high. *Leaves* broadly ovate to orbicular, 13−50×

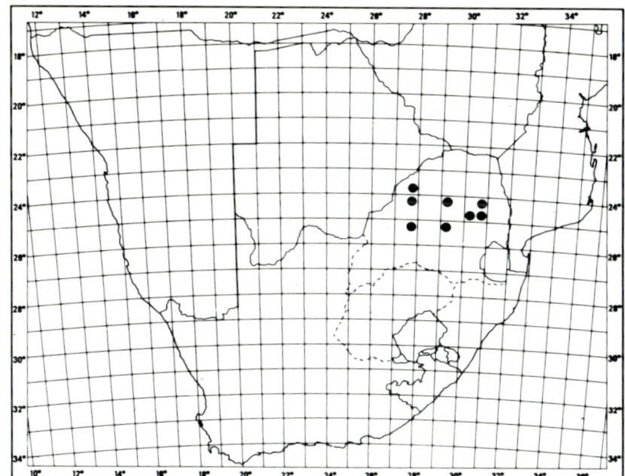

MAP 45.— **Triaspis glaucophylla**

13–52 mm, apex obtuse to emarginate, base obtuse to cordate, main vein prominent below, sometimes forming a minute apiculus, lamina glabrous to densely white-pilose, becoming glabrescent with maturity, often glaucous or slightly discolorous. *Sepals* ovate, acute, 1,5–2,5×1,0–1,5 mm. *Petals* 4,5–9(–11)× 3–5(–8) mm, lilac or bluish fading to white with age. *Stamens* with filaments 4–6 mm long; anthers ovate, 1,5–2,0 mm long. *Fruit* with wings ovate, margins entire or ± deeply emarginate at top and bottom, 19–35 mm long, usually with a few dark hairs, green becoming pale brown, sometimes tinged maroon. Fig. 20: 5 & 6.

T. glaucophylla is endemic in the Transvaal and flowers from October to April. Map 45.

Vouchers: Mönnig 63; *Van der Schijff* 7306; *Venter* 1962; *Young* A511.

2. **Triaspis hypericoides** *(DC.) Burch.*, Trav. 2: 280, 290 (Jan.–June, 1824). Syntypes: Cape, Kosi Fountain, *Burchell* 2531 (K, lecto.!); Cape, in a walk to, and on a black rocky hill under, the Kamhanni Mnts (Kuruman Hills), *Burchell* 2486 (K).

Hiraea hypericoides DC., Prodr. 1: 585 (Jan., 1824).

Scandent shrub with twining stems, sometimes small erect shrub, up to 1,5(–3) m high. *Leaves* triangular-ovate to narrowly linear-lanceolate, 13–57×4–30 mm, apex narrowly acute to obtuse, base obtuse to truncate or slightly cordate, main vein prominent below, lamina glabrous to densely white-pilose, be-

coming glabrescent with maturity. *Sepals* ovate, acute, 1,5–4×1–2 mm. *Petals* 6–11×4 –8 mm, lilac fading to white with age. *Stamens* with filaments 4–6 mm long; anthers ovate, 1,5–2,5 mm long. *Fruit* with wings orbiculate, margins entire or ± deeply emarginate at top and bottom, 25–40 mm long, glabrous or with a few dark hairs, green ripening to pale brown, often with maroon tinge.

This species occurs in S.W.A./Namibia and in central and eastern Transvaal, with a few scattered localities in Botswana and the northern Cape. It flowers from late October to May.

Three subspecies are recognized:

1a Leaf lamina canescent at all stages ....
.... 2c. subsp. *canescens*
1b Leaf lamina soon becoming glabrescent (hairs sometimes remaining on midrib beneath and on margins):
   2a Leaves 3,0–6,5 times longer than wide; occurring in Botswana and northern Cape..............
   .... 2a. subsp. *hypericoides*
   2b Leaves 1,0–3,0 times longer than wide, occurring in S.W.A./Namibia and Transvaal..........
   .... 2b. subsp. *nelsonii*

2a. subsp. **hypericoides.**

Subsp. *hypericoides* is distinguishable from subsp. *nelsonii* by its relatively narrower leaves and its distribution. Fig. 20: 4.

It should be noted that the regions in which it occurs are not yet thoroughly collected. It is possible that plants with broad leaves will also be found in Botswana or the northern Cape. Map 46.

Vouchers: *Allen* 206; *Lightfoot* 8; *Marloth* 1072 (BOL; NBG); *Ngoni* 285; *Wild & Drummond* 7248.

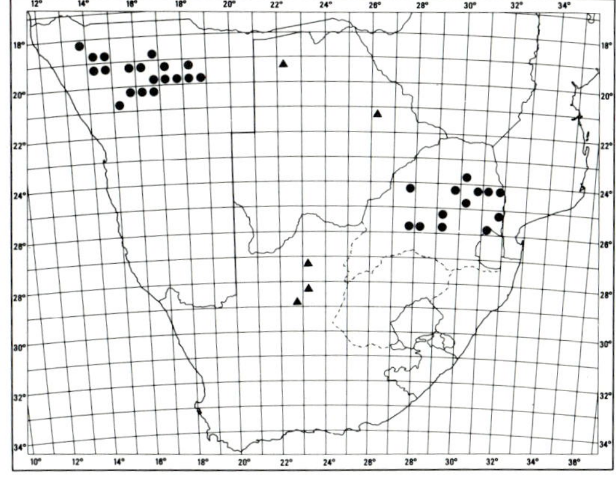

MAP 46.— ▲ **Triaspis hypericoides** subsp. **hypericoides**
      ● **Triaspis hypericoides** subsp. **nelsonii**

FIG. 20.—**Triaspis hypericoides** subsp. **nelsonii:** 1, branch with flowers and fruits, × 0,5 (*Giess* 10575 and *Le Roux* 298); 2, flower, × 2,5 (*Van Rooyen* 2169); 3, leaf, × 1 (*Giess & Smook* 10575); 4, leaf of subsp. **hypericoides,** × 1 (*Wild & Drummond* 7248). **Triaspis glaucophylla:** 5, leaf, × 1 (*Van Wyk* 5182); 6, fruit × 1 (*Germishuizen* 304).

2b. subsp. **nelsonii** (*Oliv.*) *Immelman* in Bothalia 14, 1: 79 (1982).

T. nelsonii Oliv., in Hooker's Icon. Pl. 15: 14, t. 1418 (1833); Pole Evans in Flower. Pl. S. Afr. 3: t. 109 (1923); Niedenzu in Pflanzenreich 4, 141 (Heft 91): 47 (1928); Launert in F.Z. 2: 118 (1963); F.S.W.A. 72: 2 (1968). Type: Transvaal, Pretoria, *Nelson* 511 (K, holo.; PRE!).

T. rehmannii Szyszyl., Polypet. Disc. Rehm. 2: 3 (1888). Syntypes: Transvaal, Pretoria, Aapies Poort, *Rehmann* 4195; Transvaal, Elandsrivier & drift, *Rehmann* 4944.

T. nelsonii var. *austro-occidentalis* Schinz in Vjschr. naturf. Ges. Zürich 2: 194 (1906). Syntypes: S.W.A./Namibia, Hereroland, *Schinz* 7 (Z!); Otavibei, *Dove* s.n., anno 1892 (Z!); Hereroland, Otjiraka, *Dinter* 598 (Z!).

T. nelsonii var. *glauca* Schinz, l.c. Syntypes: Transvaal, Barberton, Queen's River Valley, *Galpin* 643 (K; BOL!; NBG!; NH!; PRE!); Barberton, Moodies, *Galpin* 593.— var. *glabra* Burtt Davy, nom. superfl., Fl. Transv. 1: 51 (1926); 2: 286 (1932). Type: *Galpin* 643 (K; BOL!; NBG!; NH!; PRE!).

T. hypericoides var. *subsessilis* Burtt Davy, Fl. Transv. 1: 51 (1926); 2: 286 (1932); Niedenzu in Pflanzenreich 4, 141 (Heft 91): 48, fig. 11, H, J, (1928). Type: Transvaal, *Todd* 26 (K, holo.!).

T. thorncroftii Burtt Davy, Fl. Transv. 1: 51 (1926); 2: 286 (1932). Type: Transvaal, Tzaneen, *Rogers* 12572 (K, holo.!; BOL!; photo. at PRE!).

T. rogersii Burtt Davy, Fl. Transv. 1: 51 (1926); 2: 286 (1932). Type: Transvaal, Tzaneen, *Rogers* 12572 (K).

T. ternata Greenway & Burtt Davy, Fl. Transv. 1: 51 (1926); 2: 286 (1932). Type: Transvaal, Komatipoort, *Rogers* 22172 (K, holo.; BOL!).

T. ovata Brem. in Ann. Transv. Mus. 15: 245 (1933). Type: Transvaal, Pietersburg, Olifantsrivier, Molsgat, *Bremekamp & Schweickerdt* 450 (PRE!; PRU).

In this subspecies the leaves are variable in shape, but are no more than three times longer than wide. Fig. 20: 1, 2 & 3.

In southern Africa it is found in S.W.A./Namibia and in the Transvaal. Map 46.

In the type of *T. ternata* some of the upper leaves and inflorescences are ternate, while the lower ones are in pairs. It is probably only a rare form or mutant rather than a separate species.

Vouchers: *Craven* 561 (WIND); *Giess & Smook* 10627; *Hutchinson* 2360; *Le Roux* 631; *Smith* 1397.

2c. subsp. **canescens** (*Engl.*) *Immelman* in Bothalia 14, 1: 79 (1982).

T. canescens Engl. in Bot. Jb. 36: 249 (1905); Niedenzu in Pflanzenreich 4, 141 (Heft 91): 47 (1928). *T. nelsonii* subsp. *canescens* (Engl.) Launert in Bolm Soc. broteriana, sér. 2, 35: 32 (1961). Type: Mozambique, Ressano Garcia, *Schlechter* 11827 (B†; K, lecto.; GRA!).

Subsp. *canescens* differs from subsp. *nelsonii* only in the long hairs which persist even when the leaves are mature.

Only three specimens were seen by the author, the type and two duplicates of *Van der Schijff* 3970. The subspecies was formerly thought to occur only in Mozambique, but has since been found in the Kruger National Park.

Vouchers: *Van der Schijff* 3970 (KNP).

4219　　2. SPHEDAMNOCARPUS

by P. D. DE VILLIERS* and D. J. BOTHA**

Sphedamnocarpus *Planch. ex Benth. & Hook. f.* in Gen. Pl. 1: 256 (1862); Oliv. in F.T.A. 1: 279 (1868); Engl. in Bot. Jb. 36: 249 (1905); Niedenzu in Pflanzenreich 4,141 (Heft 93): 252 (1928); Arènes in Notul. syst., Paris 11: 97—123 (1943); Launert in F.Z. 2: 119 (1963); in F.S.W.A. 72: 1 (1968); R. A. Dyer, Gen. 1: 302 (1975). Type species: *S. pruriens* (Juss.) Szyszyl.

Scandent shrubs or climbers, deciduous; young branches often covered by an indumentum of silvery grey or golden grey trichomes; older branches often glabrous. *Leaves* opposite, ternate or rarely subopposite, simple, sericeous; petiole with 2 stalked glands; stipulate. *Inflorescence* axillary, umbellate; pedicel and peduncle sericeous; bracts narrowly ovate. *Flowers* bisexual, subactinomorphic. *Sepals* 5, free, persistent, green to greyish green. *Petals* 5, free, caducous, yellow to orange, unguiculate. *Stamens* 10; anthers pinkish, basifixed; filaments slightly flattened, connate at base. *Ovary* 3(−4)-loculed; styles 3(−4), persistent; stigmas slightly swollen. *Fruit* consisting of 3(−4) single-seeded, dorsally winged samaras.

A genus of 12 species of which one species, *S. barbosae* Launert, is restricted to Angola, one species with two subspecies occurs in southern and tropical Africa and 10 species are from Madagascar and Mauritius.

The generic name is derived from the Greek, in allusion to the sling-like fruit. In southern Africa the plants usually grow on soils derived from quartzite. The irritating hairs on the plants often cause itching.

---

* Department of Botany, Potchefstroom University for C.H.E.

** Formerly of Department of Botany, Potchefstroom University for C.H.E.; now National Botanic Gardens, Kirstenbosch.

Sphedamnocarpus pruriens *(Juss.)* *Szyszyl.,* Polypet. Disc. Rehm. 2 (1888); Niedenzu in Arb. bot. Inst. Braunsb. 6: 49 (1915); in Verz. Vorl. Akad. Braunsb. S.-Sem. 1924: 18 (1924); in Pflanzenreich 4, 141 (Heft 93): 257 (1928); Burtt Davy, Fl. Transv. 2: 284 (1932); Arènes in Notul. syst., Paris 11: 119 (1943); Launert in Bolm Soc. broteriana, sér. 2, 35: 37 (1961); in F.Z. 2: 122 (1963). Type: Natal, Port Natal, *Gueinzius* s.n. (K!).

Two subspecies are recognized:

Leaves sericeous on both surfaces ....a. subsp. *pruriens*

Leaves glabrous or nearly so, but never sericeous.......
.................................b. subsp. *galphimiifolius*

a. subsp. **pruriens.**

Acridocarpus pruriens Juss. in Annls Sci. nat., bot. 13: 272 (1840); in Archs Mus. natn. Hist. nat., Paris 3: 492 (1843); Sond. in F.C. 1: 232 (1860); *Sphedamnocarpus pruriens* var. *pruriens* (= var. *typicus* Arènes) in Notul. syst., Paris 11: 119 (1943); Launert in Bolm Soc. broteriana, sér. 2, 35: 39 (1961).

A. angolensis Juss. in Ann. Sci. nat., bot. 13: 272 (1840); in Archs Mus. natn. Hist. nat., Paris 3: 490 (1843). *S. angolensis* (Juss.) Planch. ex Oliv. in F.T.A. 1: 279 (1968); Hiern, Cat. Afr. Pl. Welw. 1: 104 (1896); Niedenzu in Arb. bot. Inst. Braunsb. 6: 48 (1916); in Verz. Vorl. Akad. Braunsb. S.-Sem. 1924: 17 (1924); in Pflanzenreich 4,141 (Heft 93): 255 (1928); Burtt Davy, Fl. Transv. 2: 284 (1932); Gossweiler & Mendonça, Carta Fitogeogr. Angola 160, 161 (1939); Arènes in Notul. syst., Paris 11: 121 (1943); Exell & Mendonça, C.F.A. 1,2: 252 (1951); White, For. Fl. N. Rhod. 183 (1962). Type: Angola, Cuanza Norte, Lopolo, *Welwitsch* 1043 (K, iso.!).

S. pulcherrimus Engl. & Gilg in Warb., Kunene-Samb.-Exped. 272 (1903). *S. angolensis* var. *pulcherrimus* (Engl. & Gilg) Niedenzu in Verz. Vorl. Akad. Braunsb. S.-Sem. 1924: 17 (1924); in Pflanzenreich 4, 141 (Heft 93) 255 (1928). Type: Angola, Biè, *Baum* 588 (B†; K, iso.!).

S. wilmsii Engl. in Bot. Jb. 36: 249 (1905); Burtt Davy, Fl. Transv. 2: 284 (1932); Arènes, Notul. syst., Paris 11: 120 (1943). *S. pruriens* forma (II) *wilmsii* (Engl.) Niedenzu in Verz. Vorl. Akad. Braunsb. S.-Sem. 1924: 18 (1924); in Pflanzenreich 4, 141 (Heft 93): 257 (1928). Type: Transvaal, cascade near Lydenburg, *Wilms* 142 (B†; K, iso., only photo. seen).

S. pruriens var. *lanceolatus* Launert in Bolm Soc. broteriana, sér. 2, 35: 42 (1961); in F.Z. 2: 122 (1963). Type: Zimbabwe, Matopos, *Rogers* 5651 (BM, holo.–PRE, photo.!).

S. pruriens var. *latifolius* Engl. in Bot. Jb. 36: 249 (1905); Launert in Bolm Soc. broteriana, sér. 2, 35: 41 (1961); in F.Z. 2: 122 (1963). *S. latifolius* (Engl.) Niedenzu in Arb. bot. Inst. Braunsb. 6: 48 (1915); in Verz. Vorl. Akad. Braunsb. S.-Sem. 1924: 17 (1924); in Pflanzenreich 4, 141 (Heft 93): 256 (1928); Burtt Davy, Fl. Transv. 2: 284 (1932); Arènes in Notul. syst., Paris 11: 120 (1943). Type: Transvaal, cascade near Lydenburg, *Wilms* 145 [B†; BM, lecto.!, vide Launert (1961)].

S. pruriens forma (I) *longipedunculatus* Niedenzu in Verz. Vorl. Akad. Braunsb. S.-Sem. 1924: 18 (1924); in Pflanzenreich 4, 141 (Heft 93): 257 (1928). Type: not indicated.

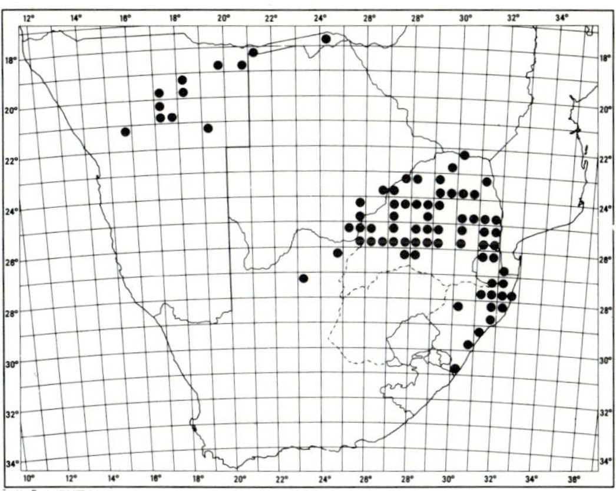

MAP 47.— **Sphedamnocarpus pruriens** subsp. **pruriens**

S. pruriens forma (III) *brevipedunculatus* Niedenzu in Verz. Vorl. Akad. Braunsb. S.-Sem. 1924: 18 (1924); in Pflanzenreich 4, 141 (Heft 93): 257 (1928). Type: not indicated.

S. pruriens var. *platypterus* Arènes in Notul. syst., Paris 11: 120 (1943). Syntypes: Mozambique, Delagoa Bay [Maputo], *Junod* 497 (P?, not traced); Lourenco Marques [Maputo], *Borle* 350 (SRGH).

Leaves opposite or whorled often with 3 leaves per whorl, lamina elliptic, ovate, obovate or lanceolate, 135×100 mm, apex acute, obtuse, mucronate or mucronulate, base cordate or obtuse, yellowish sericeous on both surfaces; petiole 2−36 mm long, with 2 stalked glands in upper half. *Stipules* reniform, ovate or obovate, sericeous, 2−10 mm long, apex obtuse or apiculate. *Inflorescence*: 2−6-flowered bracts ±3 mm long; flowers 8−15 mm in diameter. *Sepals* oblong, up to 3−5 mm. *Petals* with claw up to 1 mm long, lamina 8−12 mm long, yellow to orange. *Stamens* 2−4 mm long. *Ovary* densely sericeous; styles up to 6 mm long. *Fruit*: wing of samara 15−20×11−15 mm, sericeous to glabrescent. Fig. 21.

Found is S.W.A./Namibia and the rest of southern Africa, excluding the Orange Free State, Lesotho and Transkei. Occurs in open bush and shrubland. Map 47.

Vouchers: *Bolus* 11022; *Codd* 5195; *De Villiers* 213; *Merxmüller & Giess* 1814.

b. subsp. **galphimiifolius** *(Juss.) P. D. de Villiers & D. J. Botha,* comb. nov.

Acridocarpus galphimiifolius Juss. in Archs Mus. natn. Hist. nat., Paris 3: 491 (1843); Sond. in F. C. 1: 232 (1860). *Sphedamnocarpus galphimiifolius* (Juss.) Szyszyl., Polypet. Disc. Rehm. 2 (1888); Niedenzu in Arb. bot. Inst. Braunsb. 6: 49 (1915); in Verz. Vorl. Akad. Braunsb. S.-Sem. 1924: 18 (1924); in Pflanzenreich 4, 141 (Heft 93): 256 (1928); Burtt Davy, Fl. Transv. 2: 284 (1932); Arènes in Notul. syst., Paris 11: 118 (1943). *S. galphimiifolius*

subsp. *galphimiifolius*; Launert in Bolm Soc. broteriana, sér. 2, 35: 45 (1961); in F.Z. 2: 124 (1963). Type: Mozambique, Delagoa Bay [Maputo], *Forbes* s.n. (K, iso.!).

A. pruriens var. *laevigatus* Sond. in Linnaea 23: 22 (1850). Type: Natal, (Port Natal) *Gueinzius* 396 (K, lecto.!).

S. rehmannii Szyszyl., Polypet. Disc. Rehm. 3 (1888); Niedenzu in Pflanzenreich 4, 141 (Heft 93): 257 (1928); Burtt Davy, Fl. Transv. 2: 284 (1932); Arènes in Notul. syst., Paris 11: 117 (1943). *S. galphimiifolius* subsp. *rehmannii* (Szyszyl.) Launert in Bolm Soc. broteriana, sér. 2, 35: 45 (1961); in F.Z. 2: 124 (1963). Type: Transvaal, Houtboschberg, *Rehmann* 6390 (K, holo.!).

Triaspis transvalica Kuntze, Rev. Gen. 3: 29 (1893); Niedenzu in Pflanzenreich 4, 141 (Heft 93): 256 (1928). *S. transvalicus* (Kuntze) Burtt Davy, Fl. Transv. 1: 50 (1926); Fl. Transv. 2: 284 (1932); Arènes in Notul. syst., Paris 11: 118 (1943); Launert in Bolm Soc. broteriana, sér. 2, 35: 46 (1961); in F.Z. 2: 124 (1963). Type: Transvaal, Pretoria, *O. Kuntze* s.n. (K, holo.!).

S. rogersii Burtt Davy, Fl. Transv. 1: 50 (1926); in Fl. Transv. 2: 284 (1932); Arènes in Notul. syst., Paris 11: 118 (1943). Type: Transvaal, Pietersburg Distr., Modjadjes, *Rogers* 18041 (K, holo.!).

S. woodianus Arènes in Notul. syst., Paris 11: 118 (1943). Syntypes: Natal, Zululand, *Gerrard & M'Ken* 1788; Natal, Nanoti, *J. Medley Wood* 8921 (NH—PRE, photo.!).

This subspecies is closely related to the typical subspecies from which it differs as follows: *Lamina* 75×50 mm, trichomes restricted

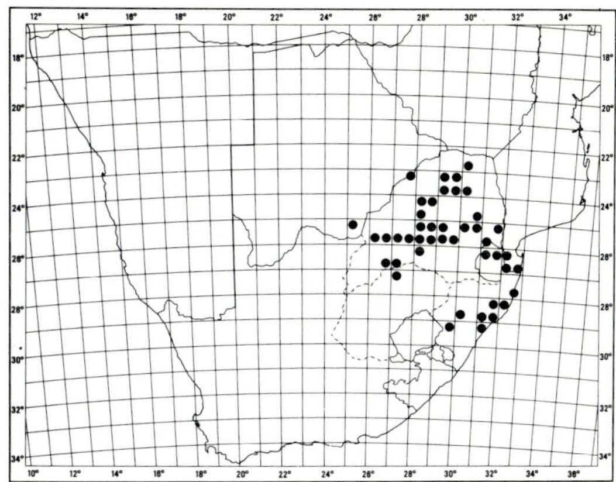

MAP 48.— **Sphedamnocarpus pruriens** subsp. **galphimiifolius**

to leaf margins, mid- and lateral veins. *Inflorescence*: bracts ±2 mm long. *Flowers* up to 25 mm in diameter. *Sepals* narrowly ovate or obovate.

Found in Botswana, Transvaal, Venda, Bophuthatswana, Swaziland, Orange Free State and Natal. Map 48.

Vouchers: *Acocks* 18739; *Botha* 2695; *Compton* 26629; *Moll* 3198; *Van Dam* 16929.

4220

3. ACRIDOCARPUS

by K. L. IMMELMAN

Acridocarpus *Guill. & Perr.* in Guill., Perr. & A. Rich., Fl. Sen. 123, t. 29 (Sept., 1831); Sonder in F.C. 1: 231 (1860), partim excl. *A. galphimiaefolius, A. pruriens* (= *Sphedamnocarpus)*; J. D. Hooker in Curtis's bot. Mag. t. 5738 (1835); Launert in F.Z. 2: 109, t. 13 (1963); Launert & Gonçalves in F.M. 32 (Malpighiaceae): 2 (1969); R. A. Dyer, Gen. 1: 302 (1975); Doorn-Kockman in Acta bot. neerl. 24 (1): 69 (1975). Type species: *A. plagiopterus* Guill. & Perr.

Anomalopteris (DC.) G. Don, Gen. Syst. 1: 634, 637 (Aug., 1831). *Heteropteris* H.B.K. sect. *Anomalopteris* DC., Prodr. 1: 592 (1824).

Perennial scandent shrubs or small trees. *Leaves* alternate or subalternate, often with glands beneath; stipules absent. *Flowers* in dense terminal racemes, 2-bracteate, bisexual, regular. *Sepals* 5, occasionally with 1—2 glands on outer surface. *Petals* 5, free, unguiculate, margins shallowly lacerate. *Stamens* 10; filaments short, free; anthers longer than filaments, opening by a short, pore-like slit at apex. *Ovary* superior, 3-locular, one loculus aborted, single pendulous ovule in each loculus; styles 2, free to the base, stigma partially lateral. *Fruit* breaking into 2 dorsally winged samaras.

The genus comprises about 30 species, all occurring in tropical and subtropical Africa, except for one each in Madagascar and New Caledonia. One species with two varieties is known from the F.S.A. area.

The generic name *Acridocarpus* means 'locust fruit' and refers to the wings on the fruits which resemble the outspread wings of a locust.

FIG. 21.—**Sphedamnocarpus pruriens** subsp. **pruriens:** 1, branch with flowers, × 0,7; 2, branch with fruits, × 0,7; 3, flower, × 3; 4, anther, × 14; 5, longitudinal section of ovary, × 4; 6, cross-section of ovary, × 4; 7, hair taken from a leaf, × 30 (all from *Rand* 431). By courtesy of the Editorial Board of *Flora Zambesiaca*.

FIG. 22.—**Acridocarpus natalitius** var. **natalitius**: 1, habit, × 0,5; 2, flower, × 2; 3, petal, × 2 (*Wells* 2142); 4, fruit, × 1 (*Leach & Bayliss* 13621).

Acridocarpus natalitius *Juss.* in Archs Mus. natn. Hist. nat., Paris 3: 486 (1843). Type: Natal, Port Natal [Durban], *Krauss* 261.

Scandent, evergreen shrub or small tree; stems rufous-pubescent when young, glabrescent. *Leaves* glabrous or rufous-pubescent, usually glabrescent with hairs often remaining on or along main veins, oblong to ovate, 30 −135(−215)×4−50(−75) mm, apex and base rounded to cuneate, midrib prominent below, dark glands sometimes present at insertion of petiole or scattered on undersurface of lamina; petiole short and stout. *Pedicels* rufous-pubescent, 12−34 mm long. *Sepals* broadly ovate, glabrous or pubescent, 2−7×1,5−5 mm. *Petals* unguiculate, spreading, concave, margins lacerated, 10−20×9−17 mm, golden yellow. *Stamens* with filaments broadly linear, 1−2 mm long; anthers 4−7 mm long. *Styles* stout, curving inward, 8−13 mm long. *Fruit* with oblong-triangular wings, 13−42×10−25 mm. Fig. 22.

Two varieties are here recognized, but with considerable reservations as they differ only in leaf width and in the fact that var. *linearifolius* does not occur in the southern part of the species range.

Leaves (13−) 15−75 mm wide; occurring throughout
    species range..............................a. var. *natalitius*
Leaves up to 13 (−15) mm wide; occurring only in
    northern Natal .... b. var. *linearifolius*

a. var. **natalitius.**

Launert in F.Z. 2: 110 (1963); Batten & Bokelmann, Wild Flow. E. Cape Prov. 94, pl. 78: 5 (1966); Coates Palgrave, Trees of Southern Africa 386 (1977).

A. natalitius var. *acuminatus* Niedenzu in Arb. bot. Inst. Braunsb. 7: 8 (1921); in Pflanzenreich 4, 141 (Heft 93): 267 (1928), p.p. excl. spec. Tanzania. Syntypes: Natal, Pondoland, *Beyrich* 93; Natal, Dumisa, near Fairfield, *Rudatis* 126; Natal, near Itafa (Ifafa?), Friedenau, *Rudatis* 352; Natal, Labendschana, Fairfield, *Schlechter* s.n.

A. natalitius var. *obtusus* Niedenzu in Arb. bot. Inst. Braunsb. 7: 8 (1921); in Pflanzenreich 4, 141 (Heft 93): 267 (1928). Syntypes: Natal, near Port Natal [Durban], *Gueinzius* 271, 392, *Plant* 26, *Schultz* 42; Natal, near Liddesdale, *Wood* 5517; Inanda, *Wood* 5517; near Durban, *Wood* 9681; near Mariannhill Trappist colony, *Landauer* 119.

A. reticulatus Burtt Davy, Fl. Transv. 1: 36 (1926). Type: Transvaal, Barberton, Crocodile Gorge, *Rogers* 23948 (BOL!).

Banisteria kraussiana Hochst. in Flora 27: 296 (1844). Type: Natal, near Natal Bay [Durban], forest margins, *Krauss* 261 (TUB?).

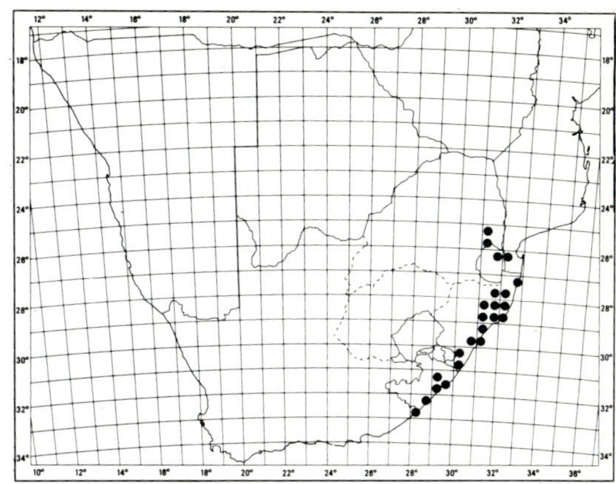

MAP 49.— **Acridocarpus natalitius** var. **natalitius**

Occurs from the eastern Transvaal through Swaziland and Natal to Transkei; also in Mozambique. Map 49.

Vouchers: *Adams* 204 (NU); *Galpin* 12088; *Leach & Bayliss* 621; *Lillicrone* 15 (NU); *Prosser* 1390; *Wells* 3446.

b. var. **linearifolius** *Launert* in Bolm Soc. broteriana, sér. 2, 35: 32 (1961); in F.Z. 2: 111 (1963); Coates Palgrave, Trees of Southern Africa 386 (1977). Type: Mozambique, Magude, Mapulanguene, *Torre* 6564 (LISC, holo.!; LAU).

A. pondoensis Engl. ex Niedenzu in Arb. bot. Inst. Braunsb. 7: 7 (1921); in Pflanzenwelt Afr. 3, 1: 830 (1950) in obs. Type: Natal, Pondoland, at edge of Egosa forest, 200−500 m, *Beyrich* 94 (B†).

In southern Africa found in Pondoland; also in Mozambique. Map 50.

Vouchers: *Pooley* 730 (NU); *Prosser* s.n.; *Strey & Moll* 3660; *Tinley* 311; *Tinley* 951 (NH, NU).

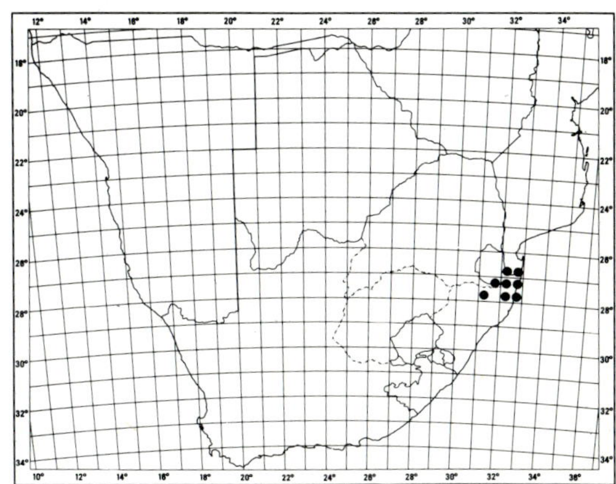

MAP 50.— **Acridocarpus natalitius** var. **linearifolius**

INDEX

---

* An asterisk signifies exotic genera and species which are not naturalized; synonyms are in italics.